SHIP IN THE WILDERNESS

SHIP IN THE WILDERNESS

Voyages of the MS "Lindblad Explorer" through the last wild places on Earth

Photographed by
JIM SNYDER

Written and illustrated by
KEITH SHACKLETON

J.M. Dent & Sons Ltd
London, Melbourne

A Gaia Original

This book was conceived, edited and designed by
Gaia Books Limited, 12 Trundle Street, London SE1 1QT

Editorial	Steve Parker Cecilia Walters
Design	Patrick Nugent Chris Meehan
Map illustrator	Ann Savage
Production	David Pearson
Direction	Joss Pearson

First published 1986 by J.M. Dent & Sons Ltd

This book is set in 11/13 Plantin Light by Capella House, Stowmarket, Suffolk
Originated and printed in Hong Kong by Mandarin Offset Marketing (H.K.) Ltd.
for J.M. Dent & Sons Ltd,
Aldine House, 33 Welbeck Street, London W1M 8LX

British Library Cataloguing in Publication Data

Shackleton, Keith
Ship in the Wilderness.
1. Scientific expeditions 2.Natural history
I. Title
508.3 QH11

ISBN 0-460-04719-1

Another book of paintings and essays by Keith Shackleton, many inspired by the Lindblad voyages:
WILDLIFE AND WILDERNESS
An Artists World
CLIVE HOLLOWAY BOOKS

To Lars-Eric Lindblad –
who dreamed her up
and made her happen

ENDPAPER '. . . As though we were truly at the world's end, and were bursting in on the birthplace of the clouds and the nesting home of the four winds . . .' Ernest Shackleton – Rough Seas, Drake Passage, Tierra del Fuego

PAGE ONE Pintado Petrel, the 'Cape Pigeon' of the early south-bound sailors.

PAGES TWO AND THREE A curtain of water cascades from the flukes of a humpbacked whale as she sounds.

BELOW The wilderness: Bismark Strait, Antarctica. The ship: *Lindblad Explorer*, 1969-84; Captain Hasse Nilsson, Master.

A small sample of *Lindblad Explorer's* wanderings across the oceans of the world.

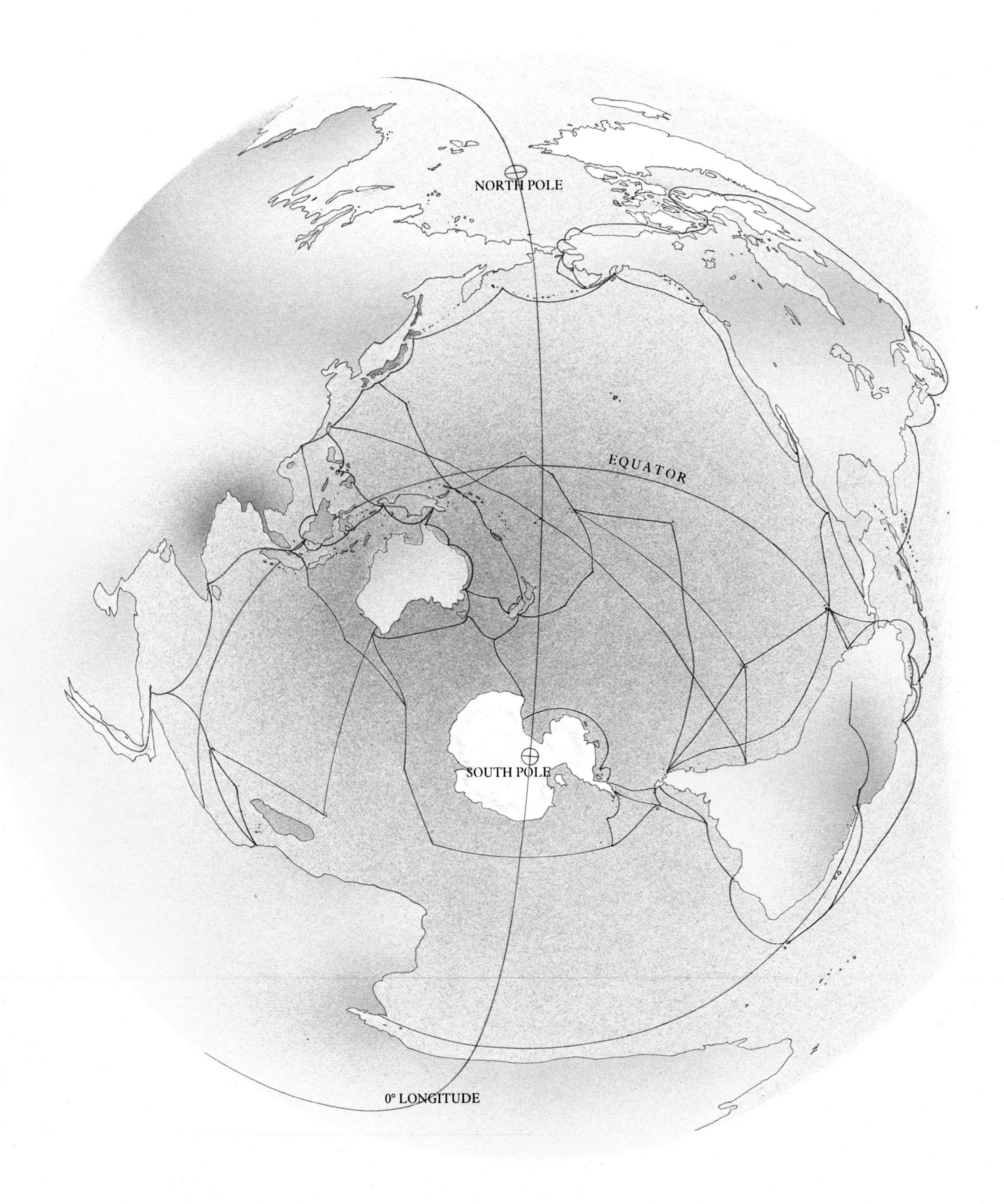

Contents

Foreword

It has been my good fortune to travel many times in the *Lindblad Explorer* with Keith Shackleton and Jim Snyder, and I am delighted to have been invited to write the foreword for their book.

It is a happy combination of talents that records an important aspect of the worldwide peregrinations of that remarkable little ship. For fifteen years she has travelled to out-of-the-way parts of the world, where passenger ships do not normally go, and has shown wild places to people who could never otherwise have seen them. And always the accent has been on the natural world and the importance of conserving it.

To see the wildlife and the fantastically beautiful scenery of the Antarctic, to gaze upon the swamp forests of the Asmat in West Irian, to savour the unbelievable colours and shapes of the Indo-Pacific coral reefs, is to become imbued with the determination to prevent them from being destroyed by the pressures of an ever-increasing human population. Great works of art have to be protected from vandals; so too does the natural world, that intricate and exquisite ecological web which is the basis of all life on earth.

It is my belief that, in showing that world to many hundreds of people down the years, *Lindblad Explorer* has played an important part in the general awareness of the need for conservation. And in producing this book Jim and Keith show us many of the wonderful wild places of the world and the people who live in them. I am delighted to have been invited to wish *Ship in the Wilderness* all success.

Sir Peter Scott

Peter Scott.

Iceberg and 'brash' ice in Penola Strait, Antarctica. *Lindblad Explorer* was a true 'ice ship'. Ice was her natural element and each year she seemed to be coming home to Antarctica for Christmas, while her romance grew with her travels in the months between.

Introduction

All of us, I think, have a foreign land that must be visited before we die. It need not be far away, but if it is its magnetism can be even stronger. All it needs is the right degree of allure: cultural, climatic, sentimental, it makes no matter. If we are lucky enough, we get there.

From about four years old, my particular siren has been Antarctica. It may have stemmed from a distant relative, of whom I was inordinately proud – Sir Ernest Shackleton. I would have gone to the great southern continent by any means, even air (it is possible) had it been the only way. But most of all I wanted to go in the time-honoured way, by sea, with the days lengthening as well as the shadows . . .

And then a week before Christmas, 1969, I left Southampton, England, bound for the ice in a brand new ship. I had been engaged as 'naturalist'. Many illustrious names were to follow me, and though I would freely admit to being the least competent scientifically, I would staunchly defend my claim to full marks for zest.

The ship was the *Lindblad Explorer*. She was not an imposing vessel, no bigger than a cross-channel ferry. You could hardly call her beautiful either, not even pretty; and from the start she carried a slight but chronic list to starboard, an imperfection we all grew to find endearing. 'There she sits, the little duck,' we used to say. 'One wing low, like the Chinese aviator.' Yet during the 15 years this book covers she stole the hearts of thousands, and re-joining her after months away was a homecoming.

I believe that ships, especially little ships, have auras like people and houses. The aura is built up of many things. Human encounter is the mainbrace; add to this the company and shared

The midnight sun of New Year's Eve plays in icicles fringing the shore of Franklin Island, in the southern Ross Sea. It also shines on Mount Erebus, 12,450 feet high. This still-active volcano dominates the whole area and has become a symbol of the heartland of Antarctic exploration. On days such as this it almost seems possible to row over there in the space of a morning – but Erebus is a good 60 miles away.

experience, bad days and good, achievement and disappointment, even moments of danger. All this rubs off on the ship, growing with the places she visits like richly-coloured beads threaded on a string. So as the little red *Lindblad Explorer* trod the oceans of the world from 1969 to 1984 her treasures lengthened, her log books piled higher and higher, until eventually she became, without any doubt, the most travelled vessel with the most exciting and varied visual record of natural wonders anywhere on the high seas.

I hope that by mentioning her list and her lack of purely aesthetic qualities, I have not presented a picture of some loveable maritime cripple. She was far from that. She was a wonder in a sea-way, though looking at her underwater lines one could never understand how she managed it. She was capricious too, her mechanical idiosyncrasies constantly testing the wizardry of her engineers.

Each successive refit brought alterations and improvements, some significant enough to change her profile. New brackets sprouted here and there to support the latest breakthroughs in electronic and navigational aids which were lavished upon her. At sea, as well as in dock, she was always adapting. At last, within the confines of her basic design, she was 'right'.

Being seriously damaged by fire off the coast of Senegal, and on her maiden voyage at that, got the *Lindblad Explorer* away to an eventful start. She suffered two bad groundings later, in the Antarctic. She sailed under five different flags and it all became grist to the mill of her personality, politics, dents and all. She became many things to many people. But if I had to make a single comment about her, I would have to say that she was a ship with 'style'.

The *Explorer's* Zodiac landing craft are the key to the whole wilderness operation. In no way do they behave like conventional small boats, yet their versatility and seaworthiness are unsurpassed. Here a Zodiac works her way – with a little human assistance – through broken pack ice to deliver mail to the British Antarctic Survey base Faraday.

Keith Shackleton

A boatload of 'SODS' leaving Grytviken, South Georgia. The Southern Ocean Drivers Society must be one of the world's most esoteric groups. Rallied under the motto 'Per Macrocystis ad Littorum' (Through Kelp to the Beach), they are responsible for the landing operations. Lamb's Navy Rum is their formally approved beverage and comes aboard by special arrangement with the Falkland Islands Company at Port Stanley.

Lars-Eric Lindblad's dream had always been to design a purpose-built ship. Years of chartering vessels that were found wanting in one way or another led to this inevitable conclusion. His concept was 'expedition cruising' and he pioneered it, fuelled by Doctor Roy Sexton of Washington DC and Captain Edwin MacDonald, late of the US Navy and 'Operation Deepfreeze'.

Expedition cruising called for small numbers of passengers with a love of wilderness and a taste for adventure, anxious to visit the remotest parts of the world. They would be well cared for and as comfortable as a small ship would allow. Each of Lars-Eric's chartered vessels had offered at least one good point to emulate. Perhaps the famous ice-ship *Magga Dan* came nearest overall; but by picking and choosing the best points of each, the broad outline of *Lindblad Explorer* was born.

She had to be small. A large vessel would be both restricted in operation and a grotesque intrusion into areas that are environmentally sensitive. The 'luxury cruise liner' image is out of the question where wildlife matters most. She would also need a shallow draught for coral reefs, shoal waters and voyages up great rivers. She required an efficient fleet of easily-launched inflatable landing craft. These would be manned where possible by scientists and lecturers who would also be responsible for running a seminar-type programme with talks, briefings and recaps. Her range had to be about 6,000 nautical miles.

Most important of all, she needed to be a polar ship, cleared for sustained operations in ice. It would have been senseless to design a 'ship for the wilderness' unable to meet the challenge of the biggest and best wilderness of all.

So her plans crystallised into a firm design – a vessel of 2,500 tons, 250 feet overall, with 46 feet beam and drawing 15 feet. She was to be powered by twin diesels developing 3,800 horsepower driving, in deference to the ice, a single variable-pitch propeller augmented by a bowthruster to give her the ultimate powers of manoeuvre. Built into her was to be a complete facility for study – lecture room with film and slide projectors,

A Sight that could fit many and widespread places; but it is in fact San Ignacio Lagoon in Baja California. Whimbrels, dunlin, willets, godwits, yellowlegs, avocets and probably quite a cast of extras are concentrated in their thousands by the high tide.

OVERLEAF Grey whales in San Ignacio Lagoon, Baja California. From November until the late spring these whales loiter off the coasts of California, the cows moving into shallow lagoons where their calves are born. Later their migration takes them northwards to summer in the Bering Sea.

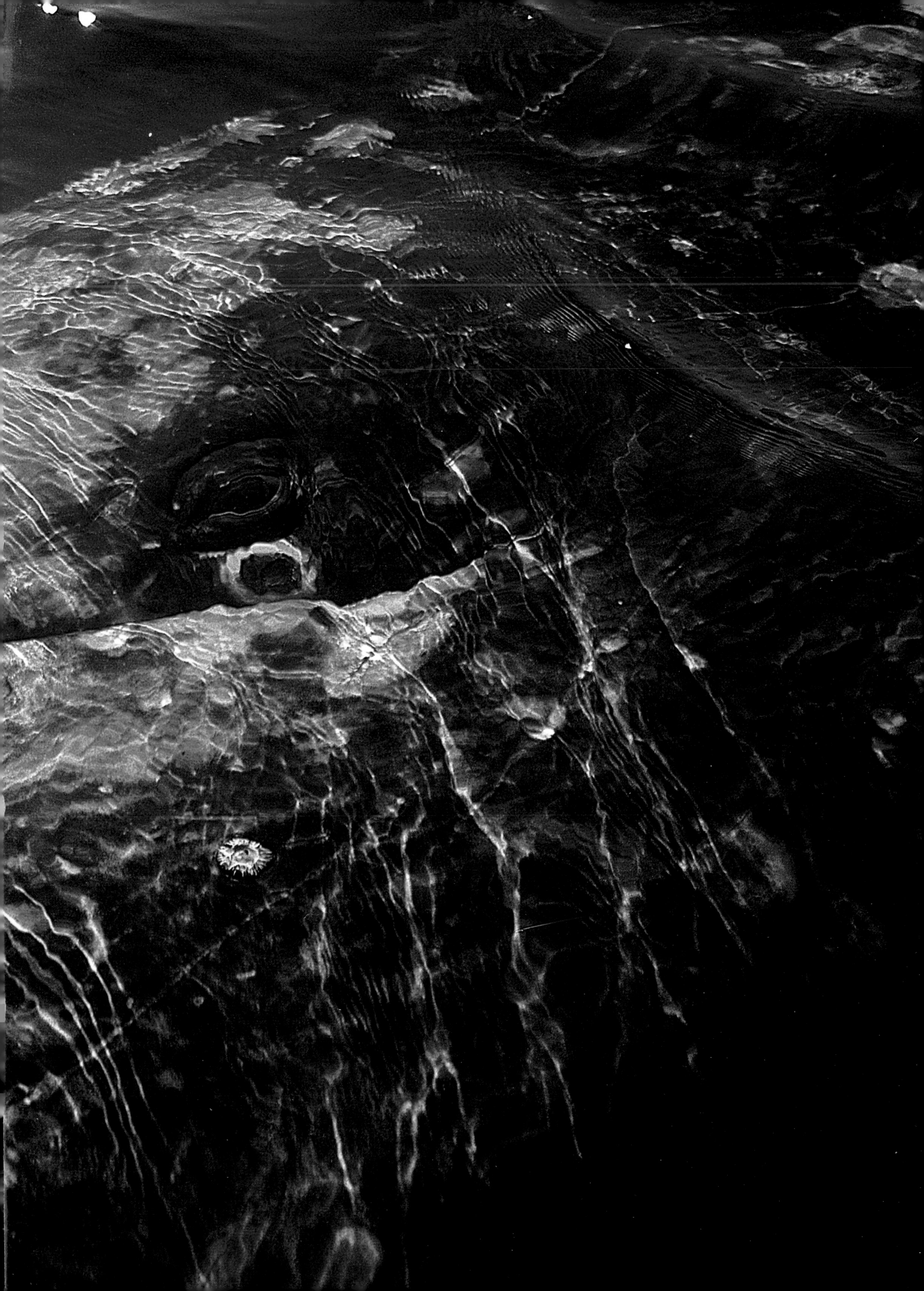

blackboard, laboratory working bench for microscopes and displays, winches aft to handle nets for plankton sampling, and everything necessary to maintain a programme of SCUBA diving. She was to meet the latest and most stringent regulations of the US Coast Guard and to be the first custom-built vessel ever to be commissioned by a travel company.

Then she happened. On 14th December 1969, she left the builder's yard in Finland on her maiden voyage – a dream come true. She was then registered in Oslo – 'a small red ship of Norwegian registry' was how the agent in Southampton described her.

She looked like a very splendid private yacht. Later on that voyage, I was ashore in Madeira and returned by taxi to the dock. She was berthed beside the mole, between the vast *Canberra* and the *Southern Cross*, looking ridiculously dwarfed like a sparkling brand-new toy.

'Which ship?' asked the taxi driver.

'The little red one.'

'The little *red* one?!' he replied with awe in his voice. 'You must be a very rich man!'

In the 15 years that followed, that little vessel was to reach farther north than any passenger ship had ever been, at 82°12′, and farthest south, too. She was to cruise over 1,300,000 nautical miles and drop her anchor over 5,000 times.

There is something special about anchoring. It suggests a ship exercising her independence and freedom to roam; indulging in personal choice on the whim of the moment; escaping from the international rip-offs of tugs and pilots and berthing alongside.

Each year saw the *Lindblad Explorer* in the Antarctic ice for the summer season, migrating north like the Arctic tern to follow the daylight. Her routes have been an ever-changing filigree of tracks covering the globe, of linking for the most part dots in the ocean, far from established shipping lanes, across the Pacific, Indian and Atlantic.

Some of these dots were strategic staging posts where international flights could repatriate her passengers and bring in new ones, but such ports of call offered the only hint of an ordered pattern. For the most part it was an oceanic amble between the poles, punctuated perhaps by the occasional voyage 2,000 miles up the Amazon to rest awhile in the biggest and most luxuriant vegetable garden on earth.

In recent years there have been imitators following in her wake. This is good; it underscores the worthiness of her concept without detraction, it complements her achievements.

Finally, to end the travel catalogue of the little red ship, in 1984 she became the first passenger vessel to navigate the

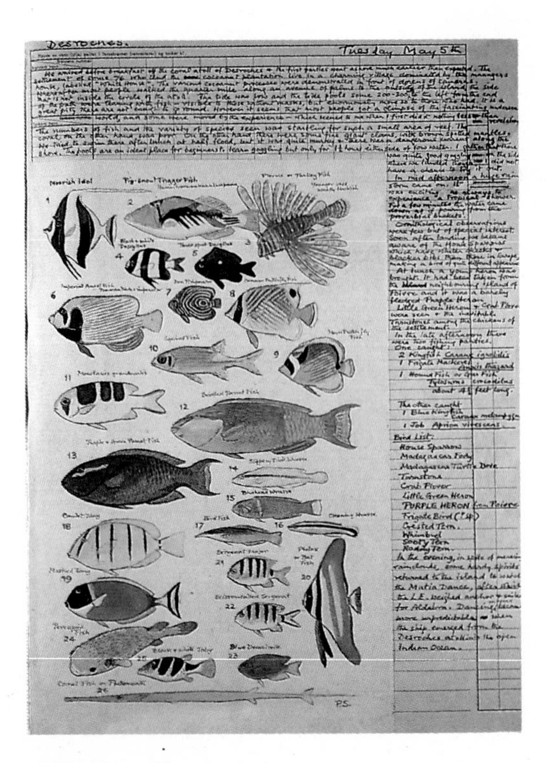

Sir Peter Scott

ABOVE RIGHT The abstract side of reality: an upturned berg in Paradise Bay, Antarctica. It is a pattern from the undersea – gutters, runnels and flooded chambers reflecting deep blue.

ABOVE FAR RIGHT A degree above freezing and a drip forms on the tip of each. Icicles, these from Franklin Island in the Ross Sea, react to light in whichever colours dominate the spectrum.

RIGHT Tridacna clams from Pinaki Island Reef show the colour effects of symbosis. Their distended, corrugated mantles are coloured by minute algae. The algae prosper in this unique protection, and in return use sunlight to produce photosynthetic sugars which help to nourish their hosts.

North-West Passage – a legendary challenge to mariners since Sir Martin Frobisher in 1576.

For my part, I have had the good fortune to work in her for some three months, on average, each year since 1969. I became quite simply her oldest inhabitant, which is to me a source of both pride and gratitude.

Many questions are asked, some by my less respectful friends at home who should know better. A typical example would be: 'What sort of nuts actually pay to go to these god-awful places?' In this respect, I have to admit that my wife is not entirely blameless. . .

After my first visit to the Falklands, I came back captivated by everything. 'I have found' I said, 'a place where I could happily settle and lay my bones. I will show you' I added, unrolling a chart of West Point Island on the kitchen table. She looked closely, her nose corrugating with distaste. 'Well' she said at last, 'this is hardly a testimonial. One end is called Cape Terrible, and at the other is Mount Misery. Are you suggesting that we convert this place marked "settlement and shearing shed" and call it Suffering Towers?' My wife, I have discovered, prefers coconut palms to ice cubes. She will jump at any chance of a voyage between the latitudes of Capricorn and Cancer, dismissing the areas marked in white as strictly for the birds – and whoever they might attract.

What sort of people were attracted? Well, they came in all shapes, sizes and dispositions. With an absolute maximum of a hundred passengers there can be no generalising; each voyage was a group of very individual individuals. While some obviously 'took off' more noticeably than others, none ever showed that corrosive apathy one meets so often ashore. In general they were people with whom it was a privilege to travel. They provided a degree of enthusiasm as nourishing as a smorgasbord, and loved it all so much that they came again and again.

They tended to be predominantly American, very well travelled, and in an age bracket that made me feel – in the early days anyway – rather youthful. Be that as it may, age seemed to have laid few limitations on them, and they were game for anything.

From the earliest chartering days the emphasis was on academic studies, so the scientific staff changed with the theatres of work. Antarctica saw a predominance of naturalists, supplemented by glaciologists, marine biologists, geologists and expedition historians. Oceanographers were welcome anywhere, as were whale experts. Anthropologists and divers marked the tropical islands and reef areas, while places like the Amazon called for botanical, local and linguistic skills. The Arctic mixture mirrored the far South with the addition of specialists on Inuit culture.

Sir Peter Scott

Marine iguanas are unique to the Galapagos but very common there. This fine specimen is from Punta Suarez on Hood Island. The world's only marine foraging lizards, they look perfectly in tune with the forbidding black lava rock and present a spectacle straight from prehistory.

A face to recall old friends, relations or even ourselves – an endearing elephant seal from Guadeloupe Island, Baja California. This female, with her huge, dark, softly-reflecting eyes and ever-dribbling nose, is of the northern species. The constant eye tears help rid her body of excess salts.

Lars-Eric Lindblad was a great believer in seeing that all this potential aboard should not be wasted, and the log books of the vessel were crammed with material that has since proved a valuable source of information in all kinds of unexpected areas – the more so because much of it covers places so seldom visited.

The log keepers were too numerous to list, and enormously varied in their approach. Some books contain copious drawings, others the written word alone. Some were illustrated with hilarious abandon by Sergio Aragonés, a cartoonist of world fame (including *Mad* magazine!). Doctor Roger Tory Peterson, Sir Peter Scott, Robert Bateman, Lyall Watson, Tom Ritchie, Sir Landsborough Thomson, Ron and Valerie Taylor, Bengt Danielsson of the Kon Tiki, and Tensing of Everest have all contributed. For some years I wrote and illustrated them myself, but my writing was as legible as a medical practitioner's and caused such problems for the long-suffering typist that when Dennis Puleston arrived, he was grabbed and never released. His is the name now synonymous with the last nine years of log books. He arrived leading an Audubon Society group in 1975 and thereafter more or less stayed for good. His presence was universally valued as the greatest asset to the ship.

With the most faithful of drawings, illustrating writing that was not just accurate but legible as well, his log books have been reproduced in facsimile after every voyage. Wherever the occasional quote appears in this text, unless another credit is given, it assuredly comes from the pen of Dennis.

What Dennis Puleston was to *Lindblad Explorer's* staff and log books, Captain Hasse Nilssen was to the ship herself. He took over command at Kristiansand in the summer of 1972, and has been her senior captain ever since. He has influenced the vessel and the way she works more than any other man, delighting in the special capabilities of his ship and using them to limits confined only by the demands of his scrupulous seamanship.

I have set down these few thoughts to give a background to this ship, so at home in the wilderness. To know something of her story will, I hope, add to the pleasure of Jim Snyder's photographs. It was the ship and her distant wanderings that made each subject so vital and so varied, and indeed possible at all. The ship is the link between every picture.

Finally, I would like this book to be a grateful tribute to Lars-Eric Lindblad, who dreamed her up and made her happen, to 'the little red ship' herself, and to all who have sailed in her.

Doctor Roger Tory Peterson

ABOVE RIGHT From Santa Cruz Island in the Galapagos comes 'Mister Jake', a great blue heron and the only professional model of his species. He was christened by the staff of the Charles Darwin Research Station and became a personal friend of the mess cook because of his appreciative attitude to left-overs.

ABOVE FAR RIGHT A red-footed booby – just one of a colony of 140,000 pairs on Tower Island, Galapagos, the largest nesting colony of this species in the world.

RIGHT Gentoo penguins take the air at Port Lockroy, Wiencke Island, Antarctica.

INTO ANTARCTICA

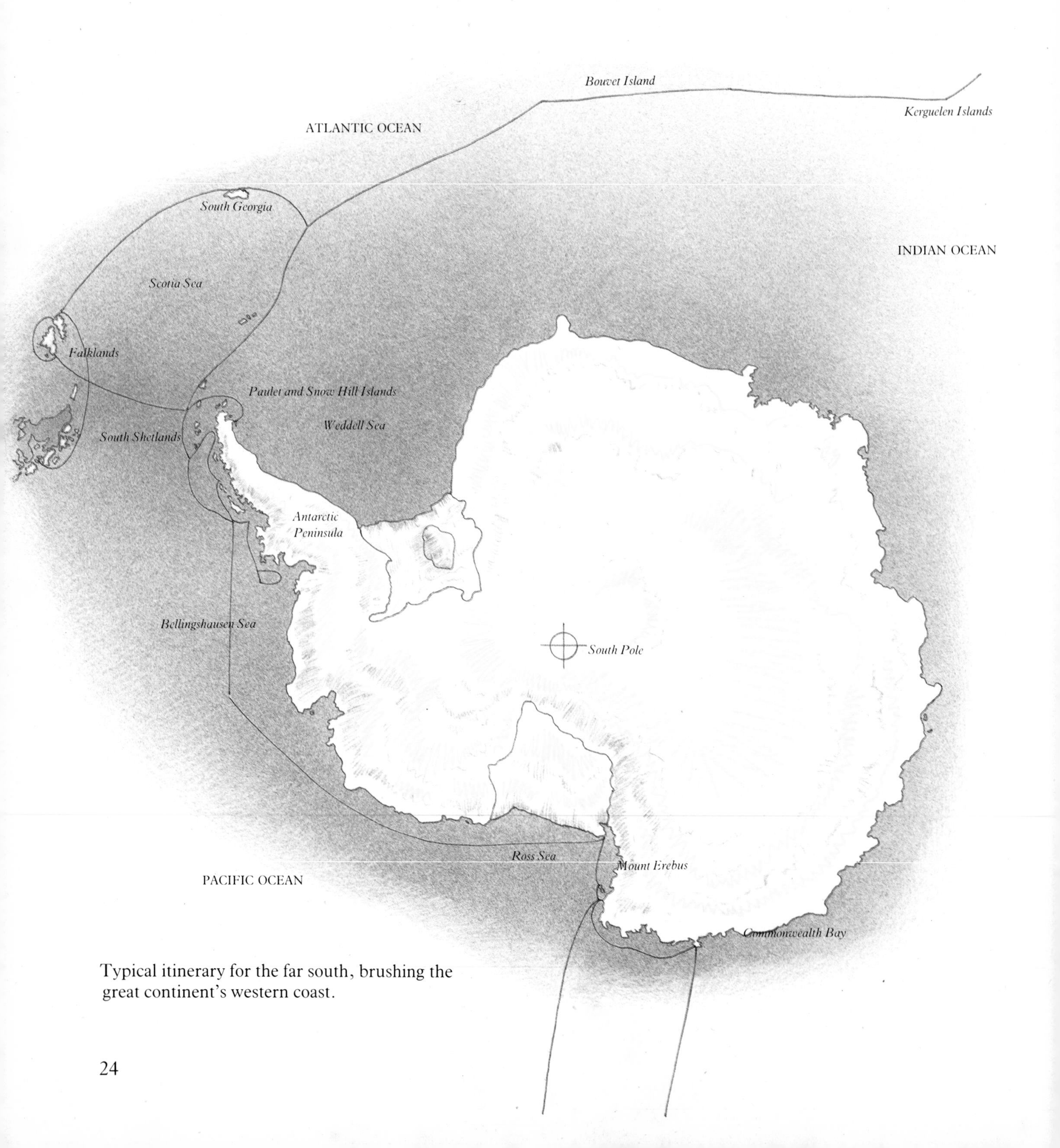

Typical itinerary for the far south, brushing the great continent's western coast.

'Great God, this is an awful place . . .' The heartfelt words of a man shortly to be overpowered, with his companions, by the most unforgiving and savage environment that the earth has to offer.

The story of Scott, his march to the South Pole, his discovery that Amundsen had preceded him and his death on the return, is one of the most moving of all human dramas – tragic yet inspirational. From it emerges a kind of yardstick by which the aura of Antarctica can be measured against human courage, endurance and sacrifice. Indeed, the two seem to be entwined like a spiritual ecosystem. One could not exist without the other. Each complements the other. In a strange way it is because of such men as Scott and Amundsen and their love-hate relationship with this fearsome continent, that a new and somewhat unexpected dimension can be discerned. We can see Antarctica in terms of what it meant to them, their hardships, what they achieved – and thereby accept it as the one wilderness that makes all others pale into insignificance.

To Sir Douglas Mawson, last great explorer of the 'heroic age', Antarctica was the 'Home of the Blizzard'. He was writing of Commonwealth Bay, south of his native Australia. To a host of subsequent mariners and scientists of many disciplines the continent has become the home of the statistic.

Antarctic statistics, like the landscape that produces them, are something to conjure with. They are practically all superlatives, often paraded in print. Here they must be paraded yet again, since without them it is hard for anyone who has not seen the place with his own eyes and felt it under his own feet, to take in the sheer enormity of it all.

In area Antarctica is about the size of the United States and Mexico combined: about 10 per cent of the earth's total land surface. Give or take a few, it covers six million square miles. An average height of around 6,000 feet above sea level makes it the highest of all the continents. It is also the coldest, and the windiest. Temperatures down to minus 88°C (minus 126.9°F) and wind speeds of over 200 miles per hour have been recorded. At the South Pole summer is a six-month day,

"Terra Nova" — January 1911.

winter a six-month night.

Antarctica holds about 90 per cent of the world's permanent supply of glacial ice, much of it over two miles thick, thereby tying up in solid state more fresh water than exists in all the lakes and rivers of the rest of the world put together. This fact invariably excites conjecture as to what might happen if all the ice was to melt, and much agonising homework has been done on the subject. The most quoted figure is that about 200 feet would be added to the mean sea level across the world. One needs little imagination or geographic perception to figure out what the new land maps would look like if the cataclysmic melting actually came to pass.

Prehistoric Antarctica was a temperate land, part of Gondwanaland – a vast southern continental complex that also took in South America, Africa, Australia and New Zealand. Green forests and flowing rivers were here millions of years ago. Then tectonic drift separated these continents and great islands and slid them into their present positions. At much the same rate the process continues today – measurable (but only just), inexorable, with land masses pulling away from each other or set on collision courses. Mementos of the primæval union that these lands once enjoyed are enshrined in the fossil-bearing rocks under Antarctica's ice. Fern leaves and grasses lie in the mountains. We see fossil trees of 150 million years ago, coal and the remains of creatures that belonged to a shirt-sleeve climate where gloves might have been worn only for protection against thorns.

Much, much later came man. Geomorphologically speaking it was in the last few minutes. Before he even saw it he gave it a name, *Terra Australis Incognita,* on the grounds that, although as yet unseen, it had to exist. It remained very incognita until 1772 when the greatest navigator of them all, Captain James Cook, filled in the picture by probing farther south than ever before. Cook and the *Endeavour* crossed the Antarctic Circle at many points, discovering new islands in the circum-polar sub-Antarctic, mapping them with dedicated accuracy, and reaching the inevitable conclusion that a vast continent really did lie to the south.

Looking at his plots and positions today, it is hard to believe that he did not, in fact, lay eyes upon Antarctica. But so firm was his integrity that he made no claims. Diary notes here and there about clouds that could almost have been mountains are easy to understand when one has been faced with the same indecision. The air can be so clear, the visibility phenomenal.

At last Cook wrote a poignant note before turning away north. It was an aside not unlike Scott's in its acceptance of events. Both men had been beguiled into heights of endeavour and resolve by a land they recognised as savage, unremitting and totally inhospitable. 'Should anyone have the fortitude to press through and achieve its discovery' wrote Cook, 'I make bold to declare that the world will derive no benefit from it . . .'

But press through they did. the sealers, then the whalers, came south to reap untouched harvests. One wonders if Cook, thoughtful and dedicated yet intensely practical, would have seen this as a 'benefit' – almost certainly, yes. Here was widened scope for discovery, maritime profit, light for lamps, furs for trade and whalebone for corsets, all offering a richer life for the human race. I think it unlikely that the role of the seals and whales in this cornucopia of bounty would have caused Cook much concern, and indeed he would have had the good grace and humility to admit that his prediction had been not entirely watertight. Man did benefit from the carnage, if only for a short span of years.

After the harvesters, others pressed through: a new generation of explorers, then the scientists, and finally, in small numbers, tourists . . .

The benefits here were of a different nature. It was now possible for people to see the continent not as a perpetual and unpredictable enemy, threatening survival, but something capable of filling the heart with pure joy and the spirit with escape.

Nobody but a moron could fail to respect the sea or any other elemental force. At the same time it is fair to say that advances since Cook have enabled us to relax our anxieties from time to time, to look around and perhaps see something of an altogether more subtle kind of benefit. In contrast to the early venturers, we can now eat the lotus. We can afford to savour life more fully; their hands were full just hanging on to it.

In summer time the human population of this vast ice-bound land stands at between 2,000 and 3,000, making it by far the world's least densely peopled region. In winter less than a quarter of these will remain. Putting it zoologically, *Homo sapiens* would be described as an 'exotic' or a 'rare vagrant'. No human being had ever been born in Antarctica until 1978. Then one or two pregnant wives of Argentinian military men were imported at the eleventh hour, with the express purpose of giving birth here and thereby establishing evidence of what is known as 'effective occupancy' – a device designed to support a claim of sovereignty.

Thinly-spread man stands here alone. The only other land animals are at the opposite end of the life scale: protozoa, nematodes, rotifers, mites and primitive insects. Not a single land vertebrate exists. There are two flowering plants, both demure little grasses growing in sheltered corners of the more northern latitudes. In the non-flowering league, lichens and mosses of surprising variety and colour decorate the rocks, their lifespan long, their growth ineffably slow. Specialised algae thrive even in the snow, spreading a green or pinkish blush. Together these mostly primitive organisms suggest a glimpse of what things might have been like in other places, when the first stirrings of life began on earth.

And here we have a paradox. The land is remarkable for its paucity of life – yet the sea is fecundity itself. Against these barren rocks and high-energy beaches, glacier tongues and ice barriers, breaks a sea among the richest in the world. Marine life from microscopic diatoms to the largest animals that ever lived flourishes in these southern waters. It is true that species are relatively few, but the populations of these hardy and successful ones seem to know no bounds. They are entwined in a web of interdependence which, robust and productive though it may be when left to its own devices, is fragile in the face of interference that might so easily throw it off balance.

The most noticeable imports from the sea are the birds. Their numbers are enormous. Though representing only a handful of species, the total number of individuals reaches an estimated 100 million. Albatrosses, petrels, storm-petrels, fulmars, cormorants, skuas, gulls and terns, sheathbills – and of course penguins.

Since Antarctica and penguins are synonymous, it should be mentioned that there are 17 species of penguin on this planet. Only four breed on the Antarctic mainland. Most of the others are sub-Antarctic, even temperate to sub-tropical. One alone breeds on the Equator. Of all the penguins, I believe the Adelie is the most 'penguin-like'; the most numerous, most quarrelsome and, some would say, the most endearing. They, and indeed all the other penguins, eat krill. The seals and the great whales eat krill. So do many of the petrels. Fish and squid eat krill, too. In fact if one were to hold up a single animal as the main shackle in the whole Antarctic food chain, krill would be it. Here, krill is the staff of life. In both size and shape it is much like a shrimp. But in proliferation it is just another of those astronomic-sounding statistics. Evidence suggests that the total population in the waters around Antarctica would amount to a weight of 650 million tons.

On 23rd June 1961 the Antarctic Treaty came into force. It had grown out of the very reasonable climate that marked International Geophysical Year, in 1957-58, and was signed by 12 nations. Since then two more have joined as 'Consultative Parties'.

The treaty's conditions are simple and could well be a blueprint for world harmony. Participants agree to pool all scientific effort, to shelve all territorial claims, to exercise no military aspirations. In short, they agree to cooperate. In 1991 it will come up for review. The world waits, with a certain trepidation, to see whether harmony will persist or whether the voices of exploitation, rivalry and jingoism will be raised.

And people ask: 'Why do you want to go to this "awful place"?'

I believe there is much concord about why people come, and Scott, despite the depth of his involvement, was no exception. Much has to do with the appeal of a wilderness, in terms of challenge and nourishment for the soul. For a painter, though, there is an extra and very obvious reason, and one that applies equally to photographers.

The landscape here is pure and intrinsic. Every shape, every line, every colour is elemental. It was moulded by natural forces alone over millions of years. Wind and water, the sky above and subterranean fire shaped it in every detail. What you see is an unsullied statement of the growing pains of the earth.

Scott knew this gentle and happier side. He sledged up one day from Cape Evans to Shackleton's 1908 hut at Cape Royds to see if the relief ship was on her way in. It was warm, sunny and calm and they sat by the penguin colony, which still exists today, to watch the birds.

'Words fail me to describe what a delightful and interesting spot this is . . .' he wrote in his diary. 'I have come to the conclusion that life in the Antarctic region can be very pleasant.'

King penguins below the Weddell Glacier, South Georgia. This group contains one young bird, whose extravagant duffel-coat of down makes him look bigger than his parents. Early sailors saw the chicks as resembling wads of oakum on the move; the 'Oakum Boys' has been their nickname ever since.

OVERLEAF A magnificent berg in the Bellingshausen Sea. In some respects icebergs can be compared with clouds. Both are free from any limitations of size, shape, or even colour.

ABOVE LEFT A young bull elephant seal from King George Island, South Shetlands. Here is deep contentment in repose, as though he is actually aware of being more attractive than his father – and intent upon relishing the situation as long as it lasts.

ABOVE A view across Anvers Island from the Neumayer Channel illustrates the simple but telling facts about Antarctica – the highest, windiest, coldest, most unremitting wilderness on earth.

OVERLEAF Shoulder to shoulder in a tight little throng, chinstrap penguins stand around like freeloaders on a weathered iceberg off Candlemas, in the South Sandwich group.

Crabeater seals hauled out on the pack ice off King George Island. The crabeater population is somewhere near 30 million, making it a favourite for the distinction of being the world's most numerous large wild mammal. The crabeater's name is a perfect oxymoron: it does not eat crabs, because there are none. Its food is the vast, dense shoals of shrimp-like crustaceans – krill.

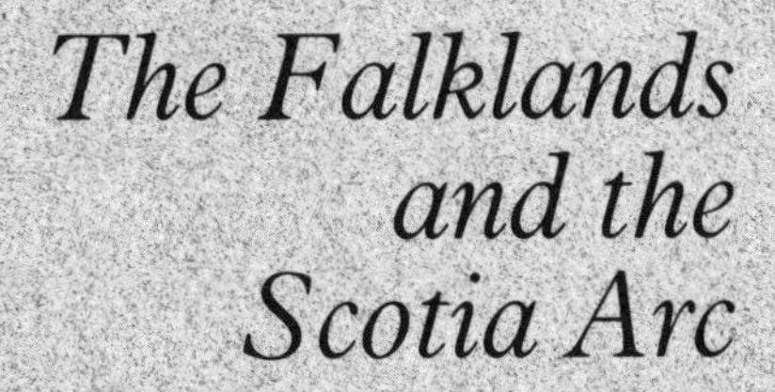

The Falklands and the Scotia Arc

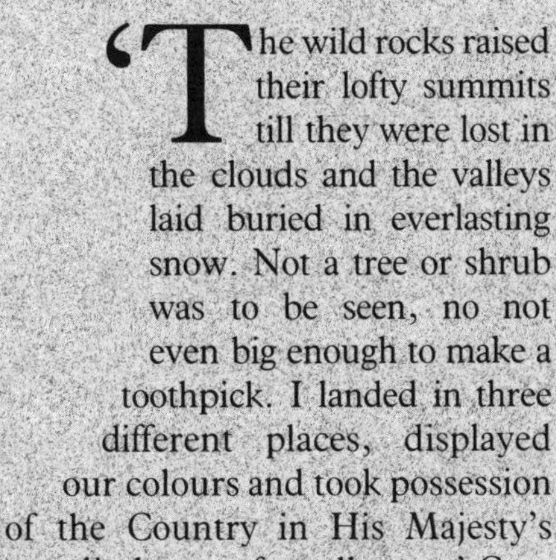

'The wild rocks raised their lofty summits till they were lost in the clouds and the valleys laid buried in everlasting snow. Not a tree or shrub was to be seen, no not even big enough to make a toothpick. I landed in three different places, displayed our colours and took possession of the Country in His Majesty's name under a discharge of small arms.' Captain James Cook, HMS *Resolution*, Possession Bay, South Georgia, 17th January 1775.

The task of moving passengers across the world to join *Lindblad Explorer* in her far-off corners always was an exercise I viewed with foreboding.

To me the modern airliner, with its unforeseen delays and cancellations, claustrophobic seating, plastic food and queuing for the loo before daybreak, adds but another appendix to the catalogue of human suffering. It can be endured only because one knows it carries a time limit and is a logistic necessity to a chosen life pattern. Drawn into myself like a miserable old turtle I dive into a paperback thriller of infinite forgettability, hoping that nobody will have the temerity to speak to me. Time is here for the killing. It is the end I want – certainly not the means.

'. . . It has been a pleasure serving you. We look forward to welcoming you aboard again . . .'

Now the end is in sight: the ship, my familiar cramped little cabin and the faces of friends. Ships have their own smells, like houses, and this one smells of contentment laced with anticipation. The moment to savour it comes when one can lean against the inside of the cabin door and shut out the world.

I mention this here because the positioning journey is inevitably long. It is not just South America, it is Tierra del Fuego. There are two principal southern ports – Ushuaia in Argentina, on the Beagle Channel; and Punta Arenas in Chile, on the Straits of Magellan. *Lindblad Explorer* is a familiar sight in both for her comings and goings into the Southern Ocean.

From the log of 23rd December 1982: '. . . hoped to sail that evening but a strong gale had arisen in the meantime and the port (Punta Arenas) was closed so the ship remained at the dock for the night.' In these latitudes, acceptance of a certain flexibility of both itineraries and timetables is as important as bunkering and victualling the vessel herself.

The following morning all was joy. Cape Froward is a romantic place to pass, being the most southerly tip of the continental New World. An interesting thought, too, that 18 months later the *Lindblad Explorer* was to pass a cable's length off Point Zenith, the most northerly, high up in the Canadian Arctic. She is very much that sort of ship.

Bound for the Scotia Arc, and immediately – albatrosses. From the outset they are there, wheeling on stiff wings over the boisterous chop of the channels and then relaxing into a slower, more rhythmic pace to match the towering white-capped mountains of ocean that march down the Drake Passage to the south of Cape Horn. Albatrosses are at one with the sea, cat-napping in the teeth of a gale, hidden suddenly by banners of spray, then vaulting upward to swing on the wind and sweep away at breakneck speed on deeply arched wings, quickly losing themselves in a running grey-green valley half a mile astern. I have often thought I would not at all mind being reincarnated as an albatross. Theirs has to be real freedom.

Drake Passage – Wandering Albatross

With Statten Island to port we head out across the north shoulder of the shallow Burdwood Bank. There are birds everywhere. Beauchene Island is an outlier of the Falklands, set far to the south of the group; a fairly low stack of flat, terraced rock and tussock-grass. In summer it is so carpeted with wall-to-wall birds you can smell it from miles to leeward. Once and only once we were able to land there. On this little dot two million rockhopper penguins were breeding – and one and a half million black-browed albatross; the flightless and the masters of flight. Congestion had reached saturation point. The din was deafening. But there they were, flipper to wing, raising their young in perfect harmony.

Beauchene, screened by a mantle of isolation, saw nothing of the recent Falklands conflict – or any other. A natural sanctuary, it has little need of legal enforcement to remain so.

Stanley, however, was different. On 26th December 1982: 'Port Stanley is no longer the quiet, isolated little town of 800 islanders. With some 4,000 military personnel stationed here now, with all their equipment and supplies and the ships to service them, it is a hive of activity. This, in addition to the masses of material left by the Argentines following their surrender, has changed the town and its environs for ever.'

Memories of the place as it used to be come flooding back. The ship would be alongside the Company Jetty – aground at low tide. Old friends on the dock. Sometimes we would walk up to Government House and there would be tea – cucumber sandwiches with the crusts cut off. We would shop for supplies at the West Store, the Kelper or the Philomel. Our exclusive organisation SODS (Southern Ocean Drivers Society, see page 13) took on the season's supply of Lamb's Navy Rum.

Everyone asked us in for tea. It was real tea, the like of which we had not tasted since leaving England. With it were home-made cakes, scones with jam and cream, Swiss rolls, sausage rolls and Lammingtons, those square sponge cakes with chocolate and coconut on them. We were without shame and snuffled about Stanley like bears in a honey factory. As 'British' as it is possible to be, the 'Kelpers' are as friendly and welcoming as any people on earth.

But events caught up with the Falkland Islands and there is no going back in any way but retrospect. Only the outer islands seemed the same – the outer islands and the weather. Sunshine and showers ('. . . if you don't like our weather, stick around a few minutes'), a buffeting wind, the scourge of hail to brighten the paintwork of the little settlements, and the smell of peat smoke that still lingers. It adds up to a climate that seems to distil energy from the air itself and wills one to walk forever over the springy turf.

South Georgia is a Dependency of the Falklands and a twoday haul by sea. '. . .On the morning of 2nd January we crossed the Antarctic Convergence.' The Convergence is a circum-polar line 20 or 30 miles wide where cold, northward-flowing Antarctic waters sink beneath the warmer waters of the sub-Antarctic. One feels the chill here as the sea temperature suddenly drops several degrees. The Convergence is really the Polar threshold. To cross it is to enter the Antarctic in the true geographic sense.

The Convergence follows a rough circle around the continent in about the high 50s of south latitude. Strangely, however, it extends a kink to embrace South Georgia while leaving the Falklands well to the north of its chilling influence. The consequence is to bestow the Falklands with the homey look and feel of the Hebrides while South Georgia is all hostility. Cook, not surprisingly, at first took the latter for a promontory of Antarctica itself until he sailed south of it, proved it an island and named its southernmost projection Cape Disappointment. I am sure he was.

As an island wilderness South Georgia is in fact anything but disappointing. To be sure it is a rough and tumble place with weather to match. Yet there are days of crystal clarity when the great Allardyce Range stands out in every detail against the blue: the high peaks, glaciers and snowfields sweeping to the ocean via tussock slopes, moraines, bogs and clear melt-water streams where king penguins congregate and elephant seals lie around in indolent heaps and belch all day. There are reindeer here, too. Brought in by Norwegian whalers early in the century, their population is still growing. The fur seals, all but dismissed for good in the sealing frenzies of the late nineteenth century, have returned with a vengeance.

In the bad old days, South Georgia's whaling stations echoed to the clank and hum of machinery, hissing steam and human invective. Giant petrels squabbled over offal. Catcher-boats fussed about and whales were so plentiful they could be taken in Cumberland Bay itself. When the catch came in the flensing knives flashed to and fro with admitted skill; the flensing plans and the sea itself was red with blood and the stench reached high into the hills.

In the early 1960s South Georgia's shore-based whaling came to an end. The stations with their boilers and storage tanks, machine shops, slipways and dry docks, mess halls, living quarters and neat little churches, have slowly fallen to pieces. Visiting trawler crews – mostly Russian – have looted and destroyed for fun. Then came the Falklands conflict and to the accumulated efforts of idle vandalism came the bomb-craters and bullet holes of warfare, the beached and broken submarine *Santa Fe,* flimsy wreckage of helicopters in the mountains and brass cartridge cases among the stones at one's feet. But across in the Bay of Isles, wandering albatrosses still sit on their nests with a brand of serenity few animals can muster, treating their human visitors to looks of disdain down long, patrician noses.

There is rare magic here: a turbulent past both begun and ended in the last hundred years. Before that only the tentative footings of explorers touched the island and left it unmarked. Sir Ernest Shackleton died aboard the *Quest* in the harbour of Grytviken, in January 1922. His grave in the whaler's cemetery is filled with tributes and some of us always walk there from the ship. Shackleton's story is too long even for précis; suffice it to say that no sailor/explorer commands more living affection and respect than he – and that from ships' companies who never knew the man, only the story and the arena of his exploits.

'6th January . . . By early afternoon the storm (65 knots) had abated and conditions enabled us to return to our original course. In the early afternoon we sighted our first tabular iceberg . . .' By the 7th, *Lindblad Explorer* was off Saunders Island in the South Sandwich group, and by the 8th, Zavodovski.

The South Sandwich group was another of James Cook's finds. In 1775 he named them for Lord Sandwich, First Lord of the Admiralty and in fact the original something-between-two-slices-of-bread Sandwich. But it was the Russian Naval Captain, Bellingshausen, who later surveyed the northernmost Traversay Islands and named them Lescov, Visokoi and Zavodovski.

Zavodovski is splendid. A classic and gorgeous volcano, 1,800 feet high, it emits great cauliflowers of smoke and steam as fickle in size and shape as they are in colour. Zavodovski himself was captain of the *Vostok* under Bellingshausen's command, but the inspiration of the British lies behind the more detailed place names that followed. The volcano herself is Mount Asphyxia. The west-facing headlands are Acrid Point and Stench Point. Noxious Bluff is to the south-west, Reek Point in the north and Pungent Point to the east. All in all one gets the message that the island could prove an olfactory experience – and reality brings no disappointment. Penguin guano, spread with unbelievable extravagance and warmed gently by vulcanism, blended with the sulphurous effluvium of rotten eggs, is the lasting memory of Zavodovski. But the penguins must have the last word.

'We had heard of an estimate of 18 million made by the Ice Patrol Vessel *Endurance* several years ago, but a cold statistic like that does not prepare one adequately for seeing such incredible numbers oneself. In addition to the multitudes on land, the sea around the ship was boiling (with birds) . . .' Other estimates go as high as 21 million. Could there be anywhere else in the whole Southern Ocean with quite such an infestation of penguins in so small a space?

A classic crowd scene: Magellanic and gentoo penguins on New Island in the Falklands. Both prefer sandy beaches to rocks for their take-offs and landings. Since man stopped killing them for oil, adult penguins have nothing to fear on land – but a host of enemies lurk in the sea, including sealions, leopard seals and killer whales. Like so many gregarious, hunted animals, they seek safety in numbers. In human terms, if one is to be taken it will 'hopefully be him and not me'.

ABOVE These Falklands gentoo penguins mass together as they go in to feed, deriving courage from closeness . . .

ABOVE RIGHT . . . and when they return they display a discernible air of relief, almost light-headedness. But with hungry sealions waiting just outside the breakers, the chances are that at least one of their number has failed to return.

OVERLEAF Multitudes beyond belief: chinstrap and macaroni penguins pressed against the volcanic slopes of Zavodovski Island in the South Sandwich group. Estimates of up to 21 million birds have been recorded for this island, making it by far the largest penguin colony on earth.

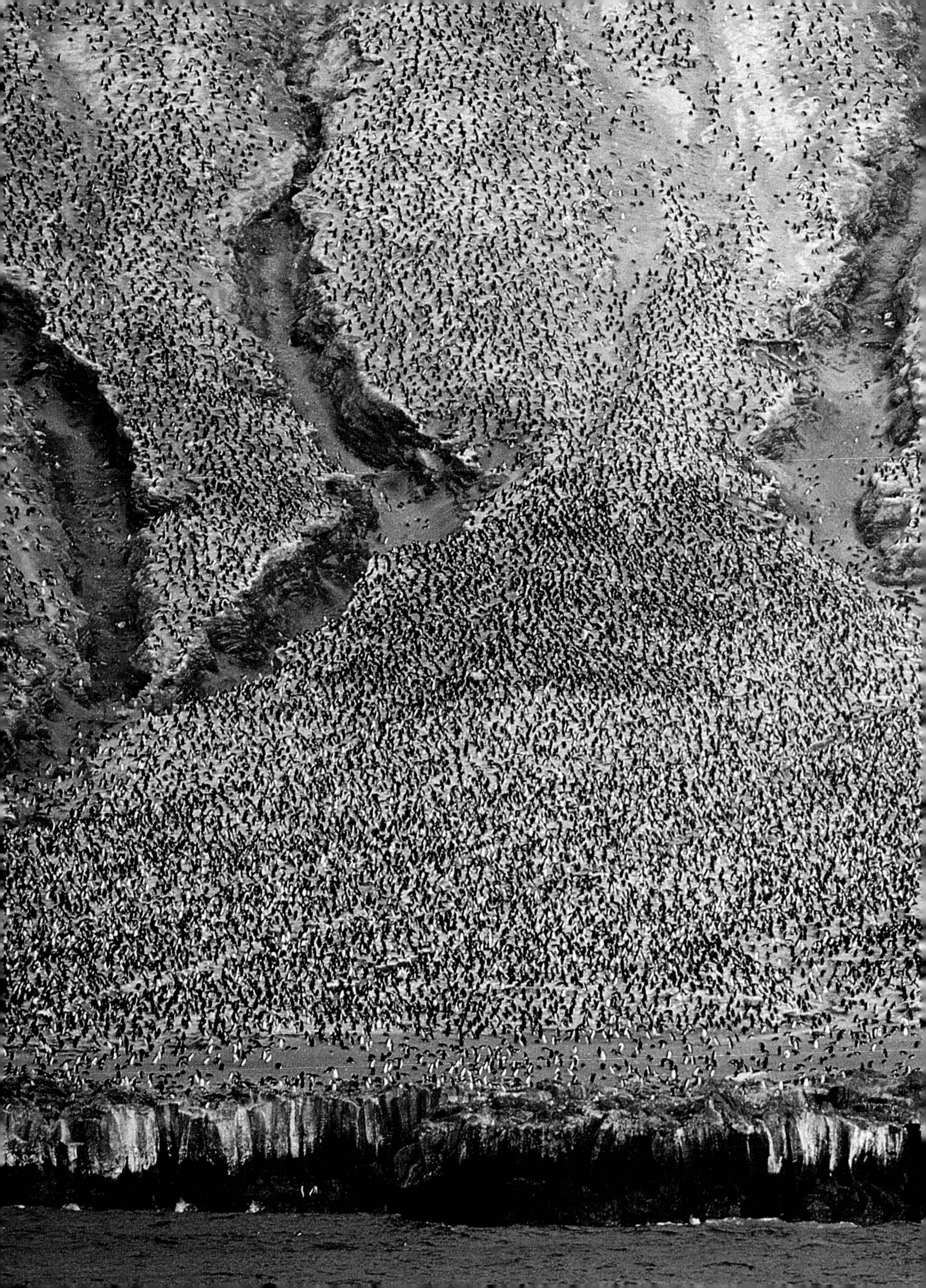

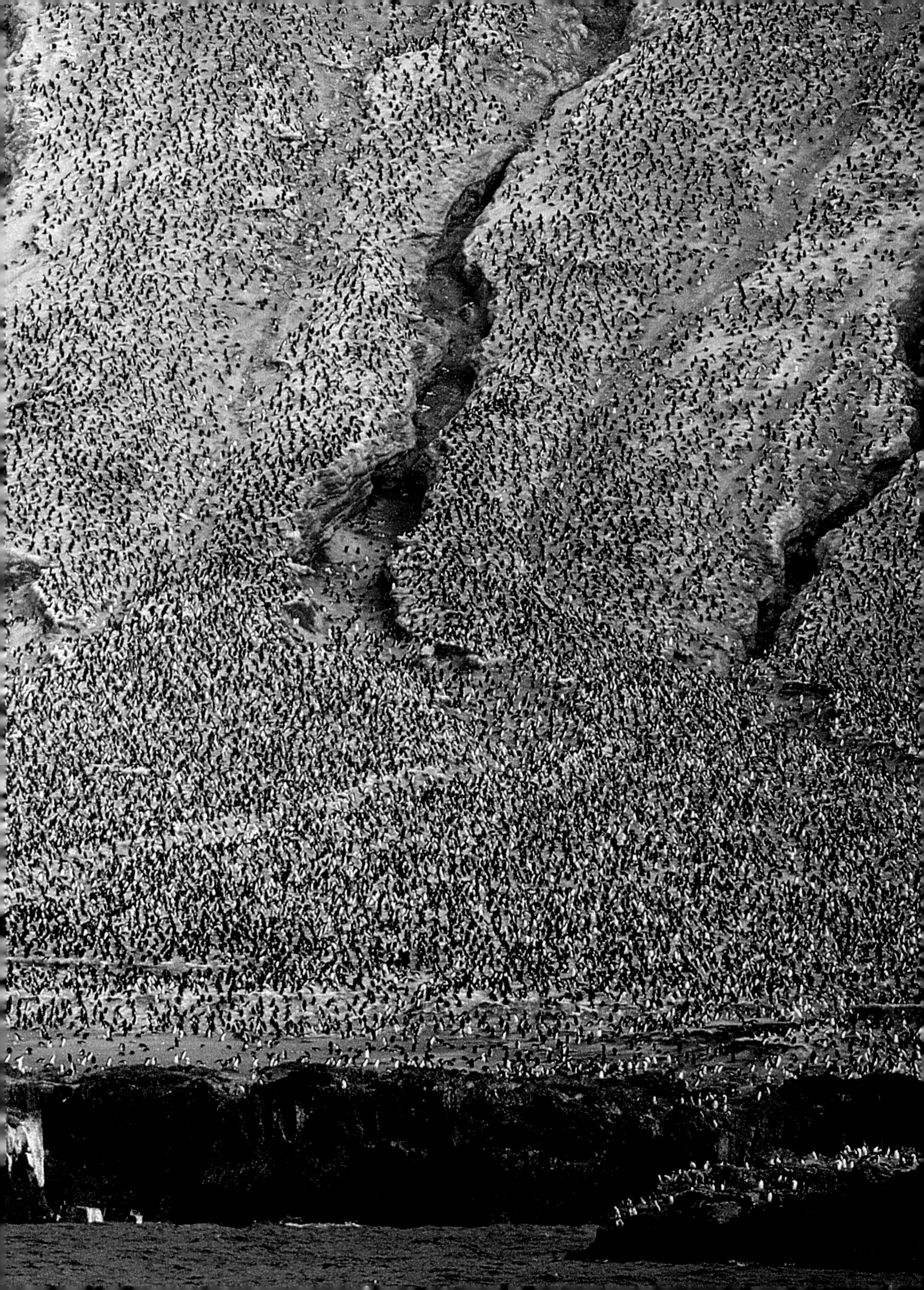

TOP 'Yankee Doodle came to town, riding on a pony, Stuck a feather in his cap and called it Macaroni.' These macaroni penguins derive their name from the 'Macaroni Dandies' – flamboyant travellers of the early eighteenth century who returned to their more prosaic English homeland having adopted the foppish styles of Europe.

ABOVE Burrow-nesting Magellanic penguins. To the Falklanders they are 'jackasses', from the raucous donkey-like braying that punctuates their waking hours.

This pair of king penguins is fresh in from the sea at Bay of Isles, South Georgia. The king is the most elegant of all penguins – nearly three feet tall, stately, long-flippered and slim, with a silver weave cape buttoned at the throat, a golden cravat and ear muffs. King penguins breed only on sub-Antarctic islands between about 45 and 55 degrees south.

ABOVE LEFT Upland goose and goslings on West Point in the Falklands. These birds are handsome, tame and on the increase, despite grazing in competition with the sheep. Their name is revered locally not only for the bird – the *Upland Goose* is also Port Stanley's largest pub.

FAR LEFT If a bird species were ever to take over the world, on the short list of candidates would certainly be the skua. This individual from Carcass Island, Falklands, is typical: a power-happy buccaneer of a bird that knows neither fear nor self-doubt.

LEFT A rock shag, or Magellanic cormorant, from West Point Island, Falklands. This species is widespread around the south coasts of Chile and Argentina as well as the Falklands.

ABOVE The black-browed albatross (this one from West Point Island, Falklands) is one of the smaller albatrosses classed as 'mollymawks', with a wing span of about eight feet.

LEFT On 7th January 1983, *Lindblad Explorer* made the third-ever seaborne landing on Saunders Island, in the South Sandwich group. She was greeted by giant petrels. 'Nellies' or 'stinkers' to the early sailors, these birds are lovely only at a great distance or on the wing. But they are just awful enough to have acquired a lot of fans, playing the part enjoyed by vultures in warmer climates.

BELOW Frenzied aggression in defence of the nest – a south polar skua at Admiralty Bay on King George Island, South Shetlands. But skuas are like people – some individuals are more aggressive than others.

ABOVE, LEFT AND RIGHT Chinstrap penguins at Waterboat Point, Paradise Bay, Antarctica. This pair is indulging in a marvellous courtship sequence of which the climax is the sky-pointing 'ecstatic' display. It is still very early in the season, the snow has not yet cleared enough to expose their rocky nesting site.

The chinstrap is a true Antarctic penguin – one of only four species that breed on the continent itself, out of a total of 17 penguin species. It is a dapper little bird, everything black or white except for the pink feet and eyes the colour of navy rum. But the best thing about the chinstrap, allowing for the inescapable 'human' comparisons struck with all penguins, is its nature – it seems akin to the very gentlest, friendliest and most considerate of mankind.

In the South Sandwich group, Zavodovski Island and Mount Asphyxia are one and the same thing. The 1,800-feet volcano stands out of the ocean, gently erupting day and night and keeping the snow at bay for much of the year. This extends the breeding season, providing rookery space on the lower slopes for millions of penguins.

BELOW A close knit colony of macaroni penguins on Zavodovski. But by far the most numerous summer populations on this island are the chinstraps.

BOTTOM Wherever penguins abound, their chief predator the leopard seal is always present. This seal has a curiously reptilian face and a mouth large enough to treat a full-grown penguin as little more than a canape. All through the summer leopard seals patrol the beach and the rim of the ice, waiting on the comings and goings of the breeding birds. Activity reaches a climax when the well-nourished but inexperienced young take to the sea.

Midsummer midnight, Wiencke Island. Though it is below the Antarctic Circle, with no direct midnight sun, a twilit hour links dusk and dawn in high summer. It can be a beautiful and tranquil period for meditation, by humans as much as gentoo penguins.

The Antarctic Peninsula

Cape Renard 2450 feet

Ever since its discovery and the painstaking surveys that followed, the British knew it as Grahamland. To the United States it was the Palmer Peninsula. To the countries nearest at hand, Chile and Argentina, it was Bernardo O'Higgins Land and San Martin Land. So in an agreed international attempt to prune confusion, it became – officially, at any rate – the Antarctic Peninsula.

Its northern tip, the Trinity, is at 60°16′ south the nearest point of the great Southern Continent to the rest of the world. It is convenient to think of this part as the 'banana belt' of the Antarctic. Aside from local catabatic gales of marrow-chilling ferocity the weather can be great.

Over the years *Lindblad Explorer* has come to the Peninsula from many directions – from the Magellan Straits, the Beagle Channel or the Falklands. Sometimes she has gone south from Cape Town via Tristan da Cunha, Gough Island and South Georgia, and sometimes from the Ross Sea on the other side of Antarctica.

Once, in 1980-81, she sailed from Singapore on a 47-day voyage ending here. Its highlights made it the most exciting voyage I remember. From leaving the Cocos (Keeling) Islands, north of Capricorn, not a soul on board – including the Captain – had ever been in these waters before. Each day brought its discoveries until we were in the familiar ice again, so tying together the two loose ends of the ship's previous polar experience with the southernmost islands of the Indian Ocean. We had anchored at Saint Paul, Amsterdam and Heard Island (then Kerguelen). There we sheltered in Christmas Bay and watched three big waterfalls cascading from a cliff-top, to be whipped back over and recycled by the gale with not a drop reaching the sea. We landed on Crozet and Bouvet and the South Sandwich group before we reached 'home ground' again in South Georgia. Such are the stepping stones to the Antarctic Peninsula; with winds and albatrosses all the way.

The final sentinels are the South Shetlands, a group of more than 20 islands forming the most westerly end of the Scotia Ridge. They were discovered in 1819 by Smith in the brig *Williams*, shaping a course nearly 500 miles to the south of Cape Horn.

All these great men who sailed down and charted this area seem to have come from Europe and North America. With the sealing trade that followed, a wealth of northern hemisphere shipping trod these waters. By 1822 nearly a hundred vessels were reported active. But sealing in the Shetlands, as elsewhere, was relatively short-lived and traffic began to dry up with the profligate over-kill. It was nearly a century before the

Melanistic Adélie Penguin Torgersen Island

fleets returned to the Southern Ocean. First came the armadas of the whaling heyday, followed thirty or forty years later by modern trawlers, factory ships and service vessels – mostly from the Soviet Block, and this time hoovering up the krill and newly-discovered stocks of fish off coasts that belong to Britain.

Smith's first sightings led to a thorough survey by Edward Bransfield, in the company of Smith and again in the *Williams*. He took possession of the largest island in the group, naming it for King George III, and also Clarence Island. Then, setting out south across the strait that now bears his name, he charted for the first time Trinity Land, this north-pointing fingertip of Antarctica itself.

In these latitudes it is always wise to be prepared for the unexpected. With voyages to the Peninsula in mind, and Cape Horn to the forefront, I searched through some old notes of my own: '. . . there stood the legendary Cape' I read, 'she of the force ten and flying spray – reflected like a great wobbling blancmange in one of Uffa Fox's "strong calms". Strings of sooty shearwaters skimmed a sea as smooth as glass . . . With a close escort of dolphins under the forefoot we headed out into the Drake Passage.'

Such a calm must be exceptionally rare but it is good to know it can happen. In the days of sail there was no way any westbound vessel under such conditions could possibly have stemmed the east-going stream through the Drake. So to pass the Horn meant wind, and by the law of averages hereabouts wind meant gales – and so the reputation of Cape Horn was assured.

Albatrosses hate a calm. They sit about on the sea like huge, incongruous bath toys, disconsolate, their 11 feet of wingspread folded six ways into a surprisingly short space. If one happens to be exactly in the path of the ship it has to move, and taking to the air is a major operation embarked upon with obvious reluctance. Remaining airborne in the calm demands a sustained flapping of the wings and to an albatross exertion of this kind is both vulgar and wholly out of kilter with its accustomed life style. Happily there is never long to wait. Deliverance begins with a play of cat's paws ruffling and darkening the water surface, which within the hour picks up to a heaving, healthy, white-capped sea and a shrieking wind to match. The birds are happy now, aloft and carving their flamboyant figures-of-eight on rested wings like skaters performing perfected repertoires before an admiring crowd.

'15th December – At sea . . . brisk wind and a beam sea . . . a lively night with most of the contents of the cabin on the deck by morning.' If albatrosses are the openers, there is no doubt what constitutes the second billing: '16th December – First ice.'

Wilson's Storm Petrel

Ice shelves abound along the base of the Peninsula. On the east side the Filchner and Larsen Ice Shelves give onto the Weddell Sea; to the west, in the Bellingshausen and Amundsen Seas, similar formations calve off their classic tabular bergs and drift north and then east. These bergs seem to know no limit in size. One was recorded in 1956, by the US Ice-breaker *Glacier*, that measured 208 miles long and 60 miles wide. Their tops are horizontal, often perfectly so. Their sides, perhaps 150 feet high, are vertical cliffs of pristine, blue-reflecting ice, often with caves and fissures in them of a deep indigo. In time they will break up along cleavage lines to form bizarre cathedral shapes complete with spires and gargoyles, or castles with battlements, floating serenely along. They roll sometimes in their disintegration to form submarines, swans, a portrait head of de Gaulle or a replica of the *Golden Hind*. Break some more and they are cottage-sized 'bergy bits' or 'growlers' and finally 'brash' the size of rowing boats, jostling in a seaway in blues and greens of unbelievable purity.

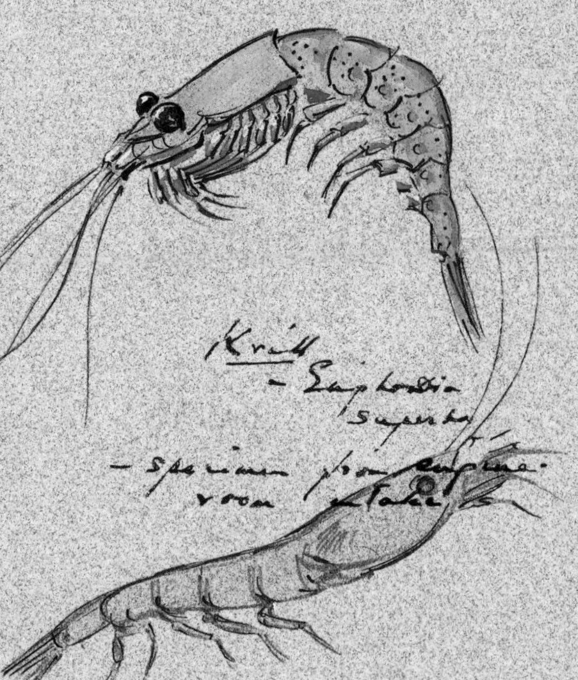

There are times when not a breath of wind stirs, when the new season's pack ice has flattened the sea by killing its swell. The parade of monumental, abstract sculpture is mirrored with such accuracy, the upside-down image as sharp as the work itself. But it is a fragile image that even the feet of a passing storm petrel may shatter into motion by dancing on the water's surface in search of food. These are days when it is good to be alive; when beauty runs out of qualifying adjectives and one is left with nothing more articulate than a sigh to express it all.

The Peninsula is not a good place for getting far south by ship – unless well offshore. Most years it is hard to force the ice as far as the Antarctic Circle at 66°30′ south. But even in the lower latitudes there is no night worth the name during the summer months.

Given fair, clear weather, the effects of a low sun are pure alchemy. The twilight colours, such a challenge to achieve in paint, touch everything with their glow. Pack ice and brash in shadow take on a fierce blue look, calm water reflects a fiery sky and around midnight, with luck, there will be a vague shadow line across the mountain massifs, slightly curved where it is cast by the sea horizon; above it a blaze of gold, orange or cherry-red on the snow peaks, lit by a still-blazing sun invisible to us at sea level below. There is no way of turning in on such nights. It is wiser to try and repay the debt of sleep at some other time.

By the small hours the sun is rising again. The curved shadow, now the colour of amethyst, is slowly descending the snow fields, throwing into sharp relief the crevasses that cut zig-zag above coastal ice-cliffs. Then with stunning suddeness, the crescendo of light seems to be everywhere despite the shallow angle of the sunrise. The transition from dusk to dawn has been completed as if with a dimmer-switch, in a matter of moments.

Ice is the ultimate sounding board for colour. Its purity is so perfect and so eager to reflect, it will receive and transmit the colours that fall on it in a way that makes them seem amplified and exaggerated. I can think of many times when such midnight spectacles have left me wondering if at last some diabolical affliction were not raising its head and that the vital link between eye and brain had been impaired – doubtless from some past over-indulgence.

But as likely as not the following day will be grey and misty, with ice crystals adrift in the air, and the only imminent problem is convincing oneself that the sights of the night before were not just a vivid dream.

Ice certainly prepares one for the Antarctic Peninsula, or seeks to do so – yet the land is still to come. Many would say that the most spectacular scenery in all the continent, and in my view this would mean in all the world, is here. Imagine the entire upper echelons of the Annapurna Himal rising from the sea, and the picture comes close. It is certainly no exaggeration: the Lemaire Channel with Mount Scott, Mount Shackleton and Mount Peary behind; Wiencke Island and its razor-backed ridge; a low sun catching Luigi Peak through a ragged, drifting hole in the clouds; and the white summit of 9,000-feet Mount Francais, with lesser peaks of the Trojan Range soaring out of the cloud blanket covering Anvers Island. These are all sights that the Peninsular offers to every point of the compass. Their counterparts could perhaps be found in the Andes, the Himalayas, the high ground of Alaska – but not the solitude. One of the stronger emotions this landscape has to offer is the sense of privilege that goes with the seeing of it.

Snow Petrel
Pagodroma nivea

The northern entrance to the Lemaire is guarded by the twin gigantic, snow-capped peaks of Cape Renard. The feature was named by de Gerlache on his *Belgica* expedition of 1897-98, and is so arresting and dominant in the scene that surveyors have made it/them the marker between the Peninsula's Graham and Danco Coasts.

Beginning with the British Grahamland Expedition of 1934, through *Operation Tabarin* in the war years, the Falkland Islands Dependencies Survey that followed, to the British Antarctic Survey of today, the striking feature of Cape Renard has been known as 'Oona's Tits'.

The question has to be asked: Who was Oona? A barmaid at the *Upland Goose*, I have heard. No, someone else will say. She was a handsome book-keeper at the Company Store in Port Stanley. Like so many romantic legends, she is becoming a fantasy of all things to all men. Of the many stories I have heard, the one I believe is that Oona was secretary to His Excellency the Governor of the Falkland Islands, that she married an upright fellow in the Diplomatic Service, and now she has several great-grandchildren and lives in Suva. Be that as it may, few ladies are born to immortality on so grand a scale – 2,450 feet of solid granite.

The story of Oona and so many like it are enshrined in the Peninsula, the spin-offs of so many voyages of exploration by courageous men in small ships from far away. Britain, Russia, the United States, Belgium, France, Norway, Sweden, Germany have all left their names on the map and their stories in history.

The Peninsula has been claimed and counter-claimed, first by Britain, then by Argentina, and then by Chile. For the time being anyway, the Antarctic Treaty agrees to put these claims literally on ice, but politics, like escaped mink, seem able to pop up anywhere.

At Hope Bay, a logical arrival point for the Antarctic Continent, now stands a large military establishment bearing the legend over the landing jetty: 'Welcome to Argentina'! 'Effective Occupancy' is still the name of the game, and it all began with the import of the Argentine pregnant ladies. Now a little school stands there for a handful of young – hopefully with a vacancy for a geography teacher!

ABOVE It is impossible to appreciate size with nothing familiar to give it scale. This curious tripod berg in the Penola Strait was measured by sextant from the bridge at 120 feet; taller than a 10-storey building.

OVERLEAF Giant icebergs drift on the current into Paradise Bay. The bay must have been named on just such a day as this.

BELOW Scores of different places, both in the Peninsula and the Ross Sea, have been suggested as the most beautiful landscape in Antarctica. Not for nothing does the Lemaire Channel find itself high on everybody's list.

BOTTOM The Gerlache Strait. During 1897-99 Adrien de Gerlache made the first survey of what is now the Danco Coast and the strait that bears his name. His ship *Belgica* became the first vessel to be trapped in the ice and suffer an enforced over-wintering.

Buttresses of black, lichen-covered rock on Torgerson Island. Lichens and mosses are the principal plant life of Antarctica. Close by is the US National Science Foundation Base of Palmer Station, on Anvers Island.

OVERLEAF A typical view across Paradise Bay. Conditions of breathless, anti-cyclonic calm are common in the Peninsula and leave one wondering if such photographs are the right way up.

A natural slipway for penguins in the Bellingshausen Sea. Some of these enormous icebergs have been at sea for months; wave action has smoothed their bases, eroding them here and there into convenient beaching ramps for penguins and seals alike.

Antarctic Highlights

One of the joys of poking about in the far South is the absence of itineraries. It is true one can list a few possibles, even optimistic probables, as potential ports of call but a lot of humility must go into it. Nothing can be guaranteed because deference is essential to weather and ice conditions when you get there. Even reaching the area at all carries some measure of uncertainty. My mother used to console herself with a ninth beatitude and quoted it often in such circumstances: 'Blessed are they that expecteth nothing, for they shall suffer no disappointment.' Thinking about it again, its applications in a wider field are enormous.

And so it happened that repeated voyages, through having been forced to seek alternatives and explore, began to build up a rewarding reference bank for the future as well as providing some stimulating new experiences for those on board at the time – or so we all hoped. Disappointment really was unknown, or perhaps transmuted into a new goal to be set aside for another time. A landing tantalisingly missed on several occasions became a must and in the end, God willing, an attainment. If there had been enough misses, the attainment became a triumph.

We have all had our favourites for our own special reasons, and some have been achieved a great deal more easily than others. For strong national interest, Hasse Nilssen and the Swedish officers and crew coveted Paulet Island in the North Weddell Sea. Paulet was named for Captain Lord George Paulet, RN, by Sir James Clark Ross on his 1839-43 expeditions, but its later brush with history is all Swedish.

Dennis Puleston's log recalls: '. . . We were interested in the stone hut built by the crew of Otto Nordenskjöld's expedition ship *Antarctic* after she was crushed and sunk by ice in 1902. During their seven-month stay on the island, one of the 22 men died. We found his grave of piled-up stones in the middle of a teeming penguin colony farther along the beach.'

For ten years we had looked with longing at this little island, rising darkly from a skirting fringe of impenetrable ice several miles wide. The north-going currents of the western Weddell Sea tend to pile up ice of every shape, from great tabular bergs calved from the Larsen Ice Shelf in the south to young broken pack of the previous year's yield, jammed tight in a polyglot jumble. Year after year we had looked across it and marvelled at the spectacle, but we could go no farther.

Then in 1981, expecting the same because ice was everywhere, we passed into the gulf and were amazed to see our island in the clear. It stood there alone, unadorned and accessible from all angles. A sea smooth enough to be lapping on the shore of a summer lake mirrored the snow-free slopes of the old volcano. The curious pepper-and-salt pattern of massed penguins covered the lower slopes, washed with the pink of droppings and vomited krill, and after the clatter of the anchor chain had subsided the raucous din of their nesting throng carried plainly over the water.

The log continues: 'Many ice-floes were grounded along the shore. In some cases we could step from one to another into deeper water, and watch the penguins swimming below us and then surfacing like small dolphins for a breath of air. There was so much constant activity it was difficult to tear ourselves away from this very special island.'

I went back to Paulet twice more. On the last occasion I climbed to the top of the extinct cinder cone, half a step downward slide for every step up. A single snow petrel circled below me like Noah's white dove in search of Ararat.

The view passed description. Paulet is virtually free of snow always. The ochre, madder, brick-red and black of minerals and the grey-green of *Usnea* lichen soar up from a slaty, ice-dotted sea. All the distant islands are totally overlain by ice and snow, yet here underfoot are the cinder screes and rock buttresses that somehow defy, even through the winter blizzards, anything more than a token mantle of white.

Before the end – The 'Antarctic' in pressure hummocks – Weddell Sea 1902 (Reconstructed from photographs by C. A. Larsen.

The Nordenskjöld story has several parts and reads like a narrative designed to prove the triumph of truth over fiction in the improbability stakes. Suffice it to say that another island of the North Weddell Sea, Snow Hill, is also involved. Here Nordenskjöld built his main expedition hut, and it still stands. We had always longed to go there but it lies south of Paulet and is beset by the same problems of the ice drift and the general rancour of the weather. In truth it is even worse than Paulet in this respect. Many times we pushed down that way and while Paulet looked difficult, Snow Hill Island seemed impossible.

Then on 21st December 1983 we tried again, pushing farther south than the *Lindblad Explorer* had ever sailed before, into the most icebound and unpredictable of seas.

'. . .We were encountering far less pack ice than normal for this time of the year, so it was decided to take advantage of this condition and attempt to land on Snow Hill . . . We were able to make landings on a stony beach and also at a farther point, from where we could search for the Nordenskjöld hut . . .'

With a timing that would have delighted Nordenskjöld himself, the party that went in to the stony beach had to leave their Zodiac because of ice. They walked along the shore and arrived at the hut just as the other party were climbing down the cliff, having crossed the island from the far side.

It came on to snow hard for the walk back over the top. A wind from the east drove the snow horizontally to form a drifted build-up in the lee of every stone that stood proud of an otherwise smooth terrain – and almost every stone was a fossil. We were amazed at the enormous numbers.

Big, beautiful bivalve clams and cockles the size of a fist; the perfectly formed ramshorn spirals of ammonites, belemnites like large-calibre rifle bullets, and whelk-like gastropods, all denizens of a shallow, temperate sea that 200 million years ago covered so much of what is now land. These beautiful fossil specimens were the greatest surprise and most lasting memory of Snow Hill Island.

Even my own very personal island goal allowed us a footing – in the end. Bouvet Island lies 54°21′ south, 3°25′ east. My diary says this about our approach to the island from the east against the prevailing westerly weather:

'14th December 1980. Back goes the glass to 965 millibars, force ten on the nose, huge seas and all afternoon hove to again in a wild white-streaked ocean with wind gusting 60 knots. First great shearwaters, several big icebergs and some nasty 'growlers' not showing on the radar for the size of the seas. This voyage is a hard slog but have now reached 15°36′ east longitude at 53°30′ south, still 424 miles to go to Bouvet . . . 15th December 1980. The 'furious fifties' were clearly not named for fun. Rolling and pitching, making nothing over the ground and a skyline like the Andes. Lots of good birds, but grey, cold and inhospitable: wandering, grey-headed, yellow-nosed, black-browed albatrosses; white-headed and kerguelen petrels, enormous numbers of diving petrels and prions.'

'16th December 1980. Mike McDowell's birthday – need I say more . . . Gave him a tee-shirt with a D flag on it (keep clear of me, I am manoeuvring with difficulty).'

'Praise Be! The glass has been rising gently but purposefully all night. We are now at 1,000 millibars (mid-morning) and at 0900 have Bouvet on the radar at 57 miles . . . Several flat icebergs, one carrying a freeload of chinstrap penguins, more southern fulmars and a single Antarctic petrel.'

'Anchored off Cape Meteor on the east side. Wind still around 17 to 20 knots but fairly sheltered. Three humpbacked whales close to the ship. Thousands of fulmars, Cape pigeons, prions and blue petrels. A party of porpoising fur seals.'

Four of us, including the birthday boy, took two Zodiacs to see if it was possible to make a landing. The beach by Cape Meteor is very narrow and swept with swell; it also has a vertical ice cliff about 300 feet high with lava rock outcrops giving onto the ice-cap itself.

'Went north along the shore inside a system of reefs and skerries – Spiess rocks – that threw up great explosions of sea. A little farther and we managed to beach both boats beyond a point called Cape Lollo, then began to wonder how many people had ever done this before . . .'

The island was discovered in 1739 by Bouvet de Lozier with the *Aigle Ann-Marie* under the French flag. He never fixed its position accurately, and though he stood off for ten days he was unable either to land or to circumnavigate it, believing it could well be a cape of a southern continent. Then in 1772

The Hut on Paulet Is.
from a contemporary drawing by Carl Skottsberg: – Nordenskjold Expedition

Cook, in the *Resolution*, sailed through 300 miles to the south to disprove this.

Cook again searched for the island in 1775, as had Captain Furneaux the previous year, but found nothing in Bouvet's supposed position. In 1808 Lindsay, in the Enderby whaler *Swan*, found the elusive island and charted it almost exactly in its true position. He too stayed a week but was unable to land.

It seems that in 1825 another Enderby vessel, *Sprightly* (Captain Norris), put men ashore – probably the first human feet. They were to be weather-bound there from 18th to 24th December, unable to get off.

In 1843, 1845, 1878 and 1898, 1916 and 1926, various ships played hide-and-seek with this island, either failing to find it altogether or finding it and slightly re-adjusting its coordinates. Only one of these was able to put men ashore. The successful visit was the Norwegian expedition of 1927 with the *Norvejia* under Captain Harald Horntvedt. They built two small huts and cached provisions for castaways; on return in 1931 the same vessel found both huts gone. But from this footing came the Norwegian annexation of Bouvet Island. From the 23rd January 1928, charts styled it as Bouvetøya (Norway).

My own notes continue: 'More forbidding than any island I have ever seen. A fearsome place. Rock and ice falls all the time and getting off was a problem. The Southern Ocean is not the best place on earth for water sports . . . Took a load of rock samples for the geologists, and safely on board again, decorated the engine room once more with mounds of sodden clothing . . .'

Bouvetøya was always my most desired island, so to stand on it was a real fulfilment. It had all started with something that caught my eye in the *Antarctic Pilot,* which must be the most evocative testimonial that can be offered to any speck on any chart. Here is how Bouvetøya is described: 'Southernmost island of the mid-Atlantic ridge, it lies approximately 1,370 miles south-west of Cape Agulhas and 1,020 miles south-east of Gough Island, and is the most isolated piece of land on the earth's surface.'

An Adélie penguin chick on Torgerson Island patiently awaits the return of parents – with food. Unmelted snow on the outside means a very warm little penguin on the inside, demonstrating the perfect insulation of the downy coat and the blubber beneath.

BELOW AND BELOW RIGHT Skuas in the Lemaire Channel. Taxonomists claim two great skuas for the far south – the Antarctic skua, and the South Polar or McCormick's skua. The latter is the greater wanderer, often passing the harsh austral winters as far north as Newfoundland and the Gulf of Alaska. Skuas have also been recorded closer to the South Pole than any other bird. Here on their home range they have become commensal with man, investigating his base camps, scientific stations and ships with a beady eye forever on the main chance. Considering what they will eat, scraps from the *Lindblad Explorer* must seem like a gastronomic breakthrough.

ABOVE LEFT AND ABOVE 'Those at the back cried "Forward!" and those at the front . . . ' — Adélie penguins on Paulet Island. The instinctive reluctance on the part of the leaders to dive in alone is a good survival strategy. One or two plunging from the icefoot are easy meat for lurking seals, while there is safety in numbers from a massed dive-in.

ABOVE 'Porpoising' is everything the word implies as an Adélie penguin breaks the calm waters of Erebus and Terror Gulf. The manoeuvre provides for a quick look round, a gulp of air, and fast travel through the water. It is also very difficult indeed to catch with a camera.

OVERLEAF At Port Lockroy on Wiencke Island, a nesting colony of blue-eyed shags. This is the only cormorant to be found in the far south and to my mind, the prettiest one in the world.

An imposing bull kerguelen fur seal from Whaler's Bay, Deception Island. A photograph taken with feeling; a male like this almost deprived us forever of the photographer. It happened on South Georgia during the aggressive competition that accompanies the rut, when beachmaster bulls are rounding up their harems. Jim was set upon after tripping and slashed in the thigh, backside and chest by a keyed-up male. Happily a lot of skilled needlework, antibiotics and three weeks' holiday restored to us a somewhat chastened Snyder.

The Ross Sea Sector

Two factors set the Ross Dependencies apart – in terms of character rather than geography – from the Antarctic Peninsula. Firstly, being a lot farther away from 'civilisation' imparts an all-pervading sense of solitude. Secondly, it is the heartland of that chapter of polar exploration remembered as the 'heroic age'.

Its significance for navigators, which certainly accounted for the exploration activity, is that by entering the Ross Sea there is the chance, indeed the probability, of sailing closer to the South Pole than anywhere else around the continent. It tends to be surprisingly ice-free in summer, and though bumping and boring is bound to be a feature at some stage, there is generally open water down to the high 70s of south latitude. Sir James Clark Ross, who christened the sea, discovered this special potential as early as 1841.

Names from that great expedition are scattered over the charts like confetti at a wedding. They sailed along the Ross Ice Shelf, a floating ice mass the size of France, marked at its northern edge by the world's biggest cliff of ice – the Ross Ice Barrier. On Ross Island, where the barrier meets the shore at its western end, stand Mount Erebus and Mount Terror, named for Ross's two ships. He spread names far and wide for himself, his ships and his men – on both sides of Antarctica. It was his well-deserved privilege. How much of it had ever been seen before through human eyes?

Lindblad Explorer first entered the Ross Sea in January 1971; she was just over a year old. My own notes begin in *The Customs House*, Hobart, a waterfront pub of enormous charm. There are odd references to schooldays in Melbourne and odd bits about Tasmania – nothing of much consequence while I waited for the ship to arrive. Then the notes mention the morning I awoke to see the *Explorer* through the bedroom window, alongside the little dock where sealing brigs had provisioned for Macquarie Island and Campbell, and where Ross himself had tied up. She had been delayed a day or two and had slipped in overnight. Previously, repeated visits to the Harbour Office for news had provided a new friend. The Harbour Master was very interested in this exciting little ship that had been hitting the headlines and he was very happy to accept a guided tour when she docked.

We sailed from Hobart, according to the notes, on 15th January at 1400 hours. A strong contingent from the Massachusetts Audubon Society was aboard, led by Bill Drury. More sophisticated cameras with outsize lenses were assembled at the taffrail, in readiness for seabirds, than I have ever seen before.

The next day was my birthday, which must account for the paucity of

notes for the 17th and 18th. But there was something about the numbers of white-capped albatrosses that kept us company on the voyage south, with all those grateful cameras burning up film.

'21st January 1970' I had written (why is it always February or even March before personal notes record the change of year?) – 'First Ice!'

'Picked up a very big iceberg on radar at 1,000 yards and lay alongside it 20 minutes later . . . heavy mist, with its top in the clouds . . . prions were flying round and the sea aglow with an aquamarine glare from the underwater ice. Heavy swell breaking on the weather side.'

'At 2300 crossed the Antarctic Circle and therefore "Glug" (that fine Scandinavian anaesthetic that is broken out on all happy occasions to make them even happier) . . . A blizzard co-incided with the Glug party to give the vessel a dusting of snow. A large number of Antarctic petrels circled us . . .'

Lower down, an interesting footnote appeared to the effect that Doctor Roger Tory Peterson, Senior Naturalist, had shown his film of the Galapagos Islands, after which a stimulating debate had developed about the domestic propagation of cactus – a sure tribute to the nimble and ever-questing mind of man!

In the years that followed, *Lindblad Explorer* made many voyages into the Ross Sea. Some have been straight cruises down via the islands from Port Bluff or Christchurch, New Zealand, or – like the first – from Tasmania. Others have taken the longer option, a voyage west from the Peninsula, often inside the polar circle in perpetual daylight. We ran through longitudes so fast there were more time changes aboard than hot dinners, to keep us in kilter with the world of people so far to the north.

There was also the International Date Line to introduce a little confusion, dictating two todays in one direction and a lost day in the other. A mental seize-up is something that comes to me easily and I always bothered about people who might lose a birthday . . .

Entry for Monday 7th January . . . 'Date Line. We have just had two Sundays (both 6th January). Both seemed like any other Sunday. Glad we missed two Mondays. Sundays are better.'

These were great voyages. On one *Lindblad Explorer* put men – and women – ashore on Peter 1st Island, only the fifth landing ever. We recorded Arctic terns that had raised their young beyond the North Polar Circle only six months before. On another we ran through the deepest depression ever recorded in the south, at 937 millibars. Byrd's figure of 942 from the 'Bear of Oakland' had stood supreme since before the war. This new low resulted in a wild welter of sea, tumbling 'growlers' and white water, but happily nothing like the ferocity such a weather system would have produced in the gentler latitudes of home. Moreover it was a following gale and despite anxious moments it made up time, the little ship fairly flying over the white-streaked sea.

One year we climbed to the crow's nest at midnight and through glasses watched Mount Erebus and the mare's-tail plume of steam from her crater, marvelling at such clarity of the air – Erebus was exactly 100 miles away.

The volcano dominates everything like an enormous cairn set there to mark the splendid human events of the far south: the courage, fortitude and sacrifice of men.

In recent years others have seen it as 'trendy', even profitable, to distort and denigrate such values, perhaps from a simple awareness of self-deficiency. But this whole vast theatre where the dramas and tragedies were played out, and where such 'historians' have never ventured, remains unchanged.

Shackleton's Hut at Cape Royds, Scott's Hut at Cape Evans, the old 'Discovery Hut' and the first hut of all, at Cape Adair, still stand, devotedly maintained by the New Zealand Antarctic Society. They still contain the gear of those expeditions between 1899 and 1917. The stores and provisions, even the magazines of the day are intact and preserved in the greatest freezer of all. Sir Douglas Mawson's Hut at Cape Dennison, in George V Land, 26° of longitude farther west, comes into the Australian Sector and is maintained to the same standards of care and devotion by the Australian Antarctic Research Programme.

The huts have an aura. To feel the effect one must be there alone or with somebody closely in tune. It is not easy to describe yet almost everyone I have spoken to has felt it, no matter how hard-boiled their demeanour. I believe it has to do with an isolated community of kindred spirits, sustained by the strongest comradeship against an ever-present background of difficulty, tension, hardship and possible death. Because no contrasting situation has followed to overlay these vibrations, they linger on. Some feel an echo of depression and foreboding but at other times there is laughter, light-hearted voices very much on the reality side of illusion, filled with expectation and hope.

1899 saw the first-ever over-wintering in Antarctica. It was at Cape Adare in the northern Ross Sea; the expedition was led by Carstens Borchgrevink, a Norwegian. His little hut on Ridley Beach still stands and *Lindblad Explorer* took in a bronze plaque in four languages, prepared in New Zealand, to tell its history to any chance passer-by in a place of perpetual solitude.

During that over-wintering their young expedition zoologist, Nicolai Hanson, died from what was described as an intestinal disease. He was buried high on the Cape in what became Antarctica's first known grave. My notes on 2nd January 1974 tell of a climb up the Cape with New Zealander Baydon Norris, a good shipmate and responsible that year for

the historic huts. We finally found the grave and reset the little metal inscription that had fallen face-down in the stones. It had been put up in 1900 by friends aboard the *Southern Cross* and it is unlikely that anyone had visited it since.

I have tried to think of special memories of the Ross Sea and there have been many. One would have to be 27th January 1972, when *Lindblad Explorer* sailed farther south than any other passenger ship, by taking advantage of an overnight gale which cleared the worst of the ice west of Cape Armitage allowing us to 77°53′ south. Five emperor penguins were there to greet us when we could break no farther.

There was the sparkling day when we berthed alongside the sea ice at Cape Royds and walked over it to Cape Evans, a 18-mile trek in blazing sunshine.

And there was the time in Shackleton's Hut when Sarah Irwin, the ship's hostess, accompanied by her own guitar, sang a song called 'South' to the tune of 'Come by the Hills'; a light and lovely female voice in what had always been a man's world.

Finally I came across some diary entries about Observation Hill and the jarrah-wood cross erected by shipmates of Scott, Wilson, Bowers, Oates and Evans: '25th January . . . climbed to the top of Observation Hill and sat for an hour, leaning against the cross . . . a white world to the south, towards the Pole. Midnight sun bright – even warm . . .'

'The outline of Mount Discovery, the Royal Society Range and all the peaks of Victoria Land stretching away to the north, in shades of saffron and amethyst. Erebus with her cloud just as Wilson sketched her . . . names of the Polar Party are still clear as ever on the cross, together with the inscription from Tennyson's *Ulysses* – "To strive, to seek, to find – and not to yield." '

Mount Erebus, on Ross Island, has to be Antarctica's most famous landmark. It was first climbed by members of Shackleton's *Nimrod* expedition in 1907-09; it dominated the view from Scott's *Terra Nova* Hut at Cape Evans (pictured overleaf); and it claimed more lives in one tragic New Zealand air disaster than Antarctica has recorded since the first human footing.

ABOVE The historic Scott's Hut from the sea, Cape Evans, Ross Island. It was built by Scott's fateful *Terra Nova* expedition of 1910-13. Since then it has been lovingly maintained by the New Zealand Antarctic Society.

RIGHT A strange presence lingers in the hut – a feeling that it is still lived in . . . provisions, equipment, magazines of the day and photographs remain in perfect preservation. The aura of Scott still dominates despite more recent occupation by the *Aurora* party supporting the Ross Sea side of Shackleton's Imperial Trans-Antarctic expedition of 1914-17.

ABOVE Penguins solemnly acknowledge Shackleton's Hut at Cape Royds. The hut was built in 1907-08 by his *Nimrod* expedition that came so close to the South Pole. It was used again by the *Aurora* party mentioned on page 86.

RIGHT Iceberg with passengers, Ross Sea. A big iceberg exerts the same magnetism for wildlife as a mid-oceanic rocky stack. Penguins hop aboard if they can; skuas, snow petrels, Antarctic petrels and gulls rest on the higher levels. Marine life abounds below. Seals and orca whales patrol among the satellite lumps of 'brash' that calve from the mother lode. There is food, or rest, or both, for them all.

LEFT Storm force ten can be a bitter wind in high latitudes, and progress slow and painful against it. Filthy weather here, off Cape Hallet in Victoria Land. It's 7th February – summer.

ABOVE Calmer conditions prevail as the *Lindblad Explorer* rests near the Antarctic coast. The ladder is down and the Zodiac landing craft drone to and fro in their ferrying of people and supplies.

ABOVE LEFT A Weddell seal on the Ross Ice Shelf. James Weddell was not just a sealer captain but a navy veteran and a great explorer. He ventured farther south (74 degrees) in 1822, into the sea that now bears his name. Weddell's sea can be a fearsome and unpredictable place; his seal, in contrast, is the friendliest and most enchanting of all.

LEFT A second Weddell portrait – right way up this time.

ABOVE Buller's albatross, Snares Island. Anyone with a love of ocean birds, and albatrosses in particular, can have a field day going south from New Zealand. This species is one of the smallest and most beautifully marked, nesting on Solander and Chatham as well as the vertical sea cliffs of Snares Island.

OVERLEAF A royal albatross – one of the 'corps élite' of pelagic birds – over Campbell Island. Life for this species follows a serene and dignified pattern. At no stage is there any semblance of haste. They even have a longer incubation period – 80 days – than any other bird.

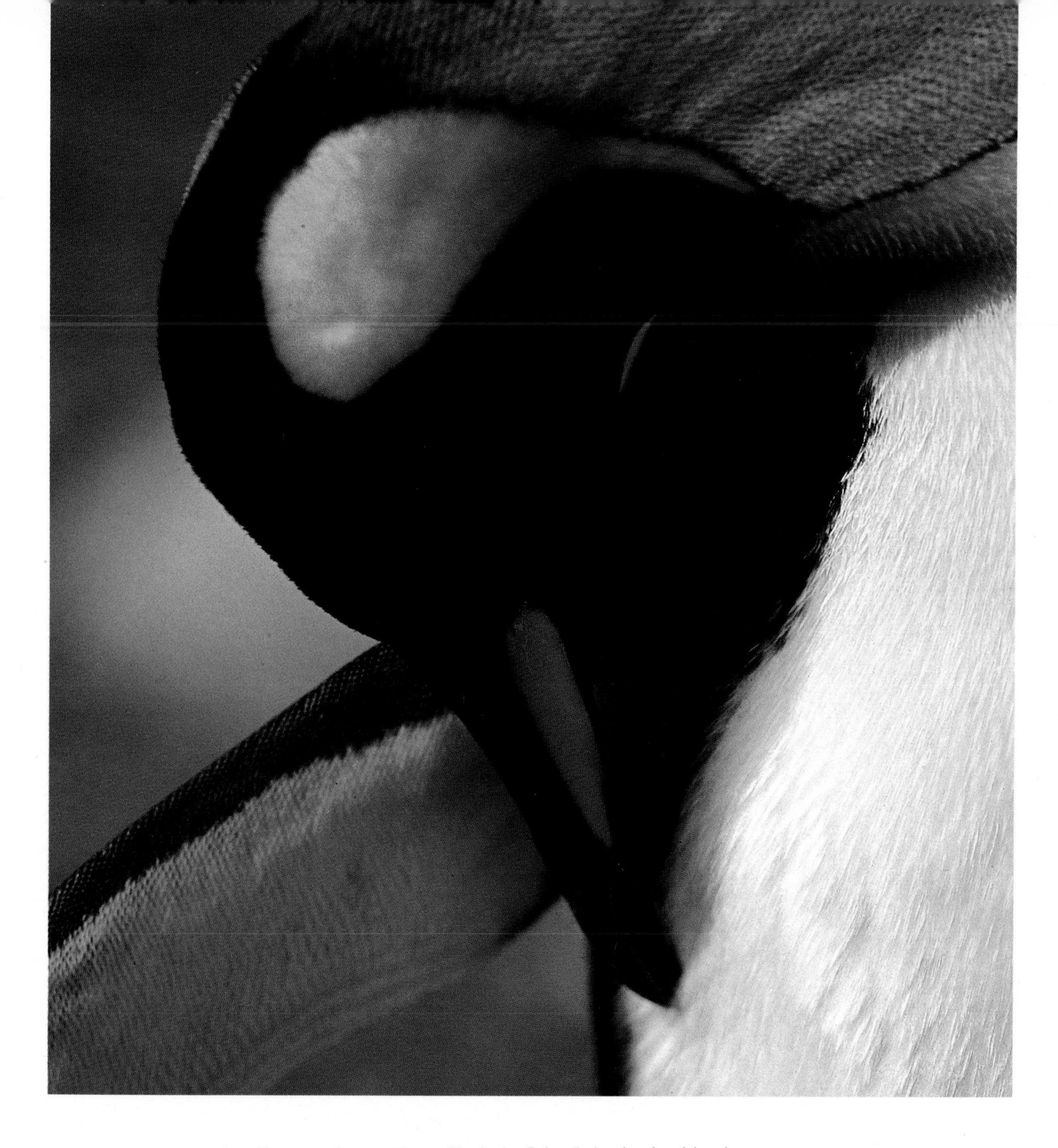

LEFT A yellow-eyed penguin on Enderby Island, in the Auckland Islands. They are secretive birds, spending much time under the tangled shrub that covers the island.

ABOVE A king penguin rules the roost on Macquarie Island. Between 1890 and 1919 an outfit shamelessly styled the 'Southern Isles Exploitation Company' all but exterminated the entire king and royal penguin populations of Macquarie. Over 150,000 birds were taken each summer, herded together and driven up long walkways to be thrust alive into the 'digesters' – great iron boilers that would render down the fat of the penguins to produce an oil suitable for soap-making. Largely due to the efforts of Sir Douglas Mawson the Tasmanian government eventually cancelled the Company's licence and the carnage ended in 1920.

ABOVE Hooker's sealions peer over the foliage on Campbell Island, in the Auckland Islands group. This species, the rarest of sealions, was probably never very numerous. Their breeding range is today confined to sub-Antarctic islands south of New Zealand.

RIGHT More Hooker's sealions, on Enderby Island. Joseph Hooker, whose name is here immortalised, was one of a distinguished group of doctor/scientists accompanying Sir James Ross on the *Erebus* and *Terror* expeditions which anchored at Enderby in 1840. Officially a botanist, he had a deep interest in all living things.

ISLANDS OF THE SOUTH PACIFIC

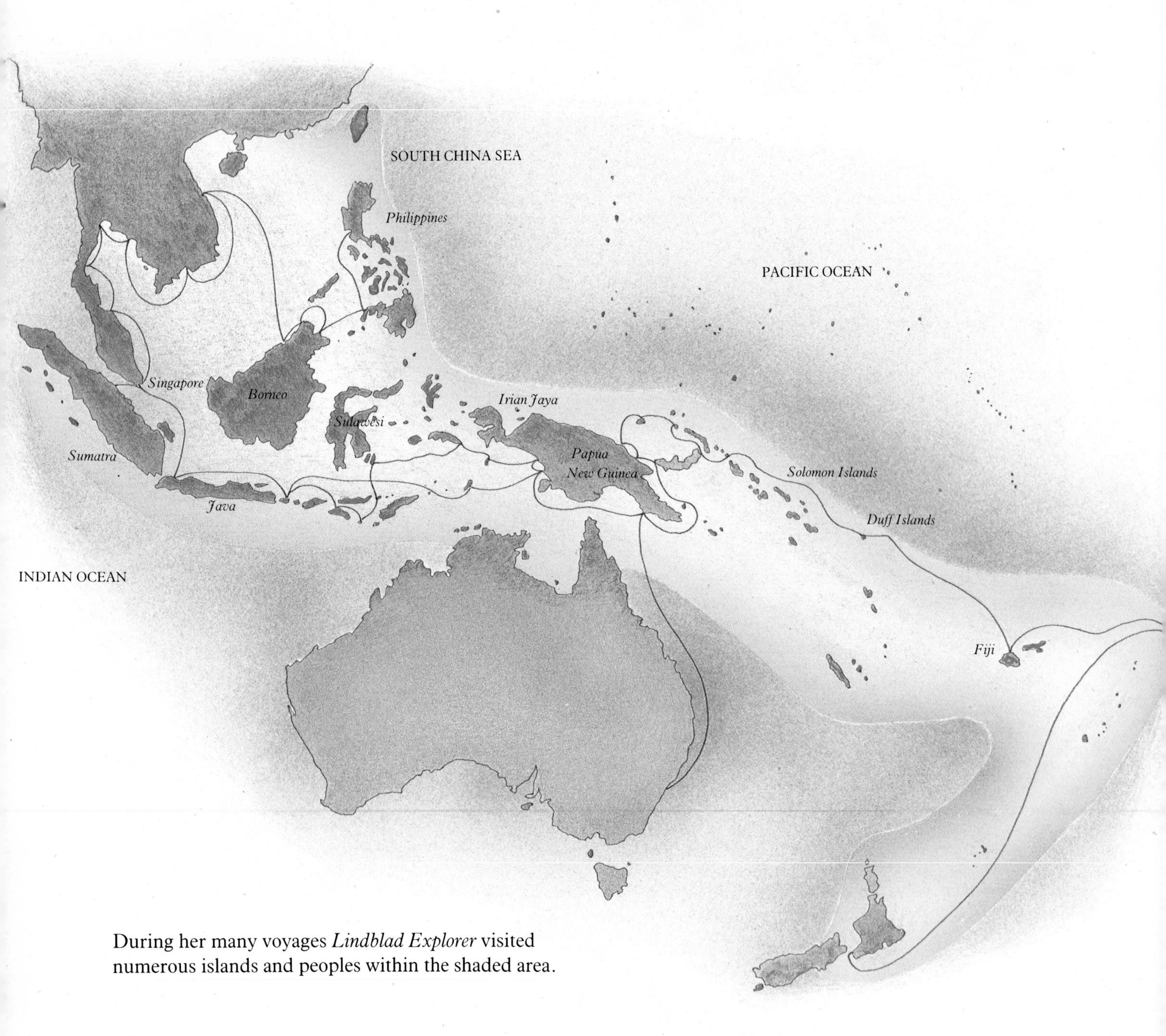

During her many voyages *Lindblad Explorer* visited numerous islands and peoples within the shaded area.

A globe offers the only way to understand how land and sea are disposed on earth. Projections, no matter how ingenious, will mislead and confuse by distortion. With a globe, one can unreeve a shoelace and trace a course from Pitcairn to Hawaii with a semblance of accuracy and more than a hint of what would be closely passed on the way.

Spin the globe and straight away the Pacific Ocean registers its enormous size. The greatest of the three oceans by far, as if in some way to redress the balance it is infested with the rash of a myriad islands. The pattern of the islands is interesting, too. They are scattered in abundance to the western (Australian) side but thinly over the eastern half that borders the New World, so thinly in fact that some globes mark an intriguing point in mid-ocean which proclaims itself farther from land than anywhere on earth. Easter Island, itself a place of deep mystery, is about the nearest patch of soil to this point.

The ethnic geography of the Pacific is also tied into the pattern of the islands. By far the biggest zone is Polynesia. Shaped like a giant ivy leaf, its point stretches out to Easter Island for its easterly extremity; the northern arm embraces Midway and Hawaii; the southern, the whole of New Zealand with Chatham Island; and the stem of the leaf runs up north-west to Nukuoro, driving a wedge between Micronesia to the north and Melanesia to the south.

Melanesia includes the whole of New Guinea, the Solomons, Vanuatu, New Caledonia and Fiji. Micronesia encompasses the Yap Islands, Guam and the Marianas, the Caroline Islands, Marshalls and Gilbert Islands.

However we split the Pacific in attempts to delineate the ethnos of the island peoples, the theatre is enormous. The area of this ocean exceeds that of all dry land; it covers a third of the surface of our planet and is greater in area than the Atlantic, Indian and Arctic Oceans combined. Statistics like these can be fascinating things and one could be excused for believing that Antarctica had used up all available superlatives, but there are more to come.

If one were to add up *all* the islands of the world that lie outside the Pacific, those within the Pacific would more than double the figure. If one were to seek out the deepest deeps of the world's oceans, they too would be found in the Pacific. A fair example is the Mariana trench: deeper by far than Everest is high, with a depth of over six nautical miles – or, for those steadfastly loyal to the most evocative of all depth units, 6,148 fathoms!

Depths like these are offset by heights. Some islands are towering volcanoes and upthrust formations of rock, whilst others lie like pancakes on the sea with imperceptible freeboard; some are low enough to be mere reefs that glow turquoise through a greater depth of blue, perhaps causing a swell to break, perhaps even showing briefly above the surface at low tide. Some have the ring form of the classic atoll and many are decorated with the forlorn bones of ships, lying there like the bodies of last night's insects below a dawn-extinguished lamp.

The clarity and warmth of the sea acts like a siren. In these latitudes, to dress in the morning is to put on a swimsuit. Each day brings a torment of indecision. Is the attraction of the white sand and palm trees, or the lure of walking around an uninhabited island barefoot to see the

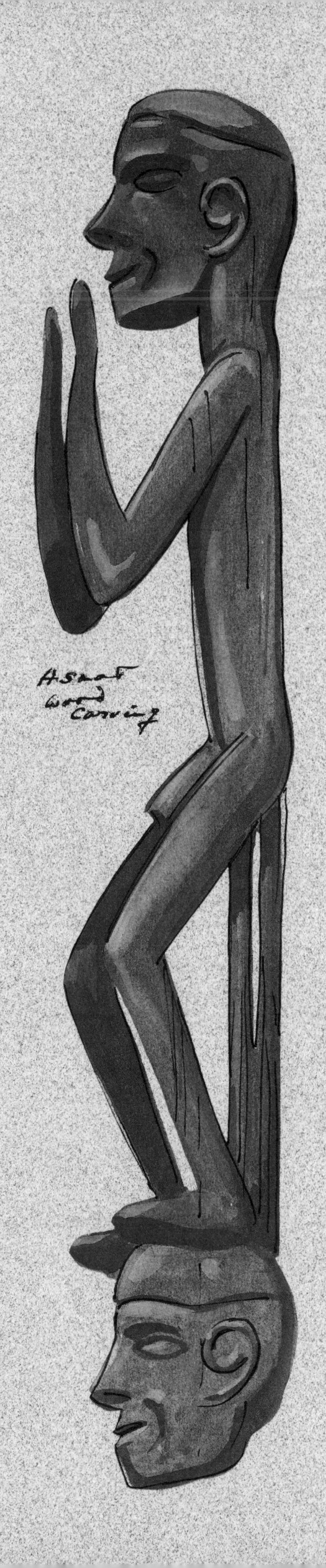

birds, more beguiling than spending the day, or most of it, grubbing about in the benthos with all those pyjama fishes? *Lindblad Explorer* was very much an underwater vessel in these parts, and much of this was due to the infectious enthusiasm and expertise of Ron and Valerie Taylor.

Had I ever been asked to name that hypothetical luxury to enrich life on a desert island, I would assuredly have asked for fins and a face-mask. Even binoculars would have come second, for on these islands the denizens above sea level will generally allow one to wander cautiously into the range of the naked eye.

To ask the desert island benefactor for more sophisticated underwater gear would have seemed extravagant, like asking Santa Claus for a Rolls Royce. But it has to be said that the SCUBA facility offered a new dimension in the ship.

The compressor for charging tanks used to putter away in the doorway of the 'Penguin Room', a laboratory-cum-lecture-theatre aft. As it filled each cylinder the local air quite naturally went in and stayed, tastefully compressed to around 200 atmospheres, until drawn upon by the happy recipient fossicking among the coral heads at some later date. Down there it provided a happy reminder of the ship because every lungful proclaimed 'Penguin Room': a special and discrete blend of diesel oil, fresh paint, distant cooking and, dare I say it, stale cigarettes. It was a smell localised to within a few feet of the Penguin Room door on the starboard side. Despite its ingredients the aroma was comforting because it spelt home. Each breath of air from those tanks, enjoyed sometimes miles away and in the most exciting of submarine situations, evoked a picture of that wretched little compressor phut-phut-phutting away by the open propped door – and brought much gratitude as well.

Recollections of the Pacific embrace the Indian Ocean, too. It was on one of the routine sailings of the early 1970s, out of Mombasa to the Seychelles, Aldabra, the Comores and Zanzibar, that *Lindblad Explorer* answered an SOS from a Formosan-registered long liner. The *Ching Fu* had run aground on Bijoutier Reef, the southernmost extremity of the Amarante Bank.

I have often thought about this episode and the ridiculous improbability of events that sometimes happen. Of all ships anywhere, ours was the only one with inflatable landing boats, and only these could have got to her. All 22 *Ching Fu* crew were taken off safely and put aboard two Royal Navy hydrographic vessels, the *Beagle* and the *Bulldog*; both these ships had sick bays, and both had answered the same call and were standing by. Such coincidences are almost unsettling.

A more recent memory: '23rd February, Deboyne Islands. In the early morning the ship entered the great circle of reefs enclosing the Deboyne Islands group. To our surprise we saw a small uncharted island on the eastern tip of the reef, on our port hand as we entered the channel. It was a strip of sand about 150 yards long covered with low scrub and five coconut palms' . . . Since it was not on the chart, we decided to name it Lex Island and plant a company flag on it. This led to many proposals, including declaring it a separate nation, imposing a 200-mile limit, joining the International Whaling Commission . . . Later in the morning, a Zodiac made an official landing there and a flag was duly planted. Leading government officials are yet to be appointed pending the drafting of a constitution.'

It is interesting to reflect that Lex Island, according to the charts, was one fathom under water at high water springtides and would have been noted as a high point in a complicated reef system. It is an exciting thing to be in at so swift a birth of a new land, already supporting trees and collecting jetsam on its tide line; a land that has not been born like Surtsey, off Iceland, by fire and earthquake and all the attending violence of volcanic midwifery.

Writing this way about the Pacific, jumping here and there thousands of miles at a time, is like trying to distil two or three random lines dealing with half the world and expecting it to be conclusive. Take the globe again between two opposing fingers; the Indo-Pacific is more than half, and this little ship has covered almost every part within the embrace. Memories come thick and fast – from the guano islands off Chile to the Horn of Africa, from the tempestuous sub-Antarctic seas to the sleepy, palm-shaded, colourfully-peopled islands that need a Bengt Danielsson to unravel the interplay of migratory mankind.

Like drawing numbers from a hat, I have taken some final random log book extracts from this ocean, spread in dates from 1971 through to 1983.

Gorong Islands. Tom Ritchie: 'The morning found us at a village famous for its magic. After some welcome dances, a few of the locals demonstrated glass-eating and self-mutilation with needles while showing no evidence of pain. Well, to each his own . . .'

10th November. At sea. Jim Snyder: 'An overcast day of rest highlighted by Bengt Danielsson's spirited description of the *Bounty's* aftermath. As late afternoon approached, the sun struggled through the clouds to grace our arrival as we circled Pitcairn Island and dropped anchor in Bounty Bay.'

17th June. Asmat. Christopher Powell: 'Approaching the mouth of the river the Zodiacs, with startling rapidity, were completely surrounded by no less than 21 war canoes. This was a totally different and almost frightening experience in the very midst of some 150 armed men in full warpaint and regalia. Furthermore, they clearly demonstrated that they could elicit a far greater speed from their canoes than our fully-loaded Zodiacs.'

And finally, Dennis Puleston again: '13th March. Palawan Island. In the morning we cruised southwards down Palawan's west coast (south-west Philippines) until we found an anchorage off the village of Sabang. Fifteen minutes in the Zodiacs brought us to a fine white beach

set between high, forested hills. At the northern end of the beach a shallow river was flowing into the sea. This is the Saint Paul subterranean river. We embarked in several Zodiacs and outrigger canoes and entered underground using Coleman lanterns and torches to probe the darkness . . .'

Dennis goes on to mention the stalactites of every imaginable form and colour; the thousands of bats of several species roosting from the high vaulted ceilings; the glossy swiftlets nesting; the voyage of over a mile before turning back to emerge, finally and suddenly, and with eyes adjusted for darkness, into the noon-day glare.

Those of the Southern Ocean Drivers Society who happened to be present were moved to call a special meeting on the beach. The toast was to the Pacific Ocean which had given us yet another new experience: that of being called upon to navigate our little vessels, for the very first time, into the bowels of the earth – and, happily, out again.

The tropical Pacific, and parts of the Indian Ocean too, are strewn with atolls and tiny islands like this. They seem to have no freeboard but float on rafts of coral and sand that glow through the shallow water in tones of turquoise and eau-de-Nil. This could be any of a thousand island guesses. It is in fact Pileni, in the Santa Cruz Islands.

LEFT A maiden of Map Island, Yap. Happily there are still places where 'unspoiled' seems to be the only adjective. The inhabitants have managed to hold at bay the revolution of nylon and plastic, and one likes to think that their own inclinations have played some part in it. If room still exists for the unashamed romantic, it must be somewhere in the Pacific. The spell that cast itself on the *Bounty* mutineers is still alive and well – and living in this ocean.

BELOW 'Coral reef with Outrigger, Sail and Islanders'. One of the Yap Islands' outriggers designed to fish the inshore reef. An intricate harmony with the sea and the wind is the birthright of all these island peoples; moreover, it begins to show itself from the cradle.

LEFT Like an illustration from a child's geography book, with the caption 'This is a volcano', Tinakula dominates the Santa Cruz Islands in the Solomons. Its almost perfectly symmetrical cone is surmounted by a self-generated cumulo-nimbus of smoke and steam, towering far above the normal weather build-up. By night its eruptions are an awe-inspiring spectacle.

ABOVE A view from the crater rim of Mount Yasua, the great volcano of Tanna in the New Hebrides (now Vanuatu). As we peer over, the ground trembles underfoot and a deep rumble erupts from the bowels of the earth. Boulders, in the slow-motion that goes only with giant size, rise spinning hundreds of feet into the air, hang for a moment and drop back into the cauldron.

BELOW To the people of Anuta Island, in the Solomons, *Lindblad Explorer* represented a happening of far greater consequence than the everyday local volcano. Before the anchor was down every canoe on the island was alongside. Those unable to scramble aboard one, just swam out.

RIGHT The long, elegant and lavishly carved dugouts of the Asmat. The Asmat region is the shallow estuarine landscape of southern West Irian – the old Dutch New Guinea. Even today these arcane, stone-age people are dangerous, unpredictable and still engaged in tribal vendetta and ritual head-hunting. It was here that Michael Rockefeller was lost in circumstances of sinister mystery.

OVERLEAF An aerial view of mud-stained waters around a settlement on the Kawawari River, Papua New Guinea. With rainfall at around 200 inches a year, flooding is a part of life in the Sepik Provinces.

ABOVE An Asmat warrior of Biwar Laut, West Irian. Historically his people were the fiercest warriors of New Guinea. A legend claims that the first Asmat was not only a warrior but a wood-carver, so he chose the strongest trees from which to carve his own army and lead it into battle. But when the work was finished he saw that his men were no more than statues. Undeterred, he sought out the greatest tree of all, chiselled out a drum from its heart – and beat his wooden warriors into life . . .

RIGHT Another warlike face, from Pirien, in West Irian. Nose ornaments of the Asmat vary from the carefully sculptured bones of ancestors to whorled plates fashioned from pearl oysters, clams or nautilus shell.

LEFT A burial cave at Kete Kesu village, Toraja, in Sulawesi. At first, the dead of Toraja are deemed only to be sick. There follows a series of rituals after which the body may be acknowledged as officially dead and allowed to lie in state. Those who were of high rank are wrapped, embalmed and kept in a central room of the house until the body becomes desiccated. The ensuing funeral feast may be delayed for months, even years; only after that can the mortal remains be laid out in Toraja's limestone caves.

ABOVE In Toraja these life-sized wooden effigies, each one dressed in clothing from the deceased, guard the burial tombs of Lemo against the spirits of the underworld.

LEFT A Torajan house interior in Siguntu, Sulawesi. The Tongkonan is a traditional dwelling often beautifully decorated with buffalo heads and bird motifs. Buffalos are highly revered as symbols of wealth and status and are never used as draft animals. In Toraja, ploughs are pulled by men.

ABOVE Modelling Savunese *ikat* at Seba on Savu Island. A year's work could go into garments like these. To produce ikat each strand must be painstakingly tied off and dyed by hand, a strand at a time, before it can be woven into the unique cotton cloth for which Savu Island is famous.

THE ENCHANTED ISLES

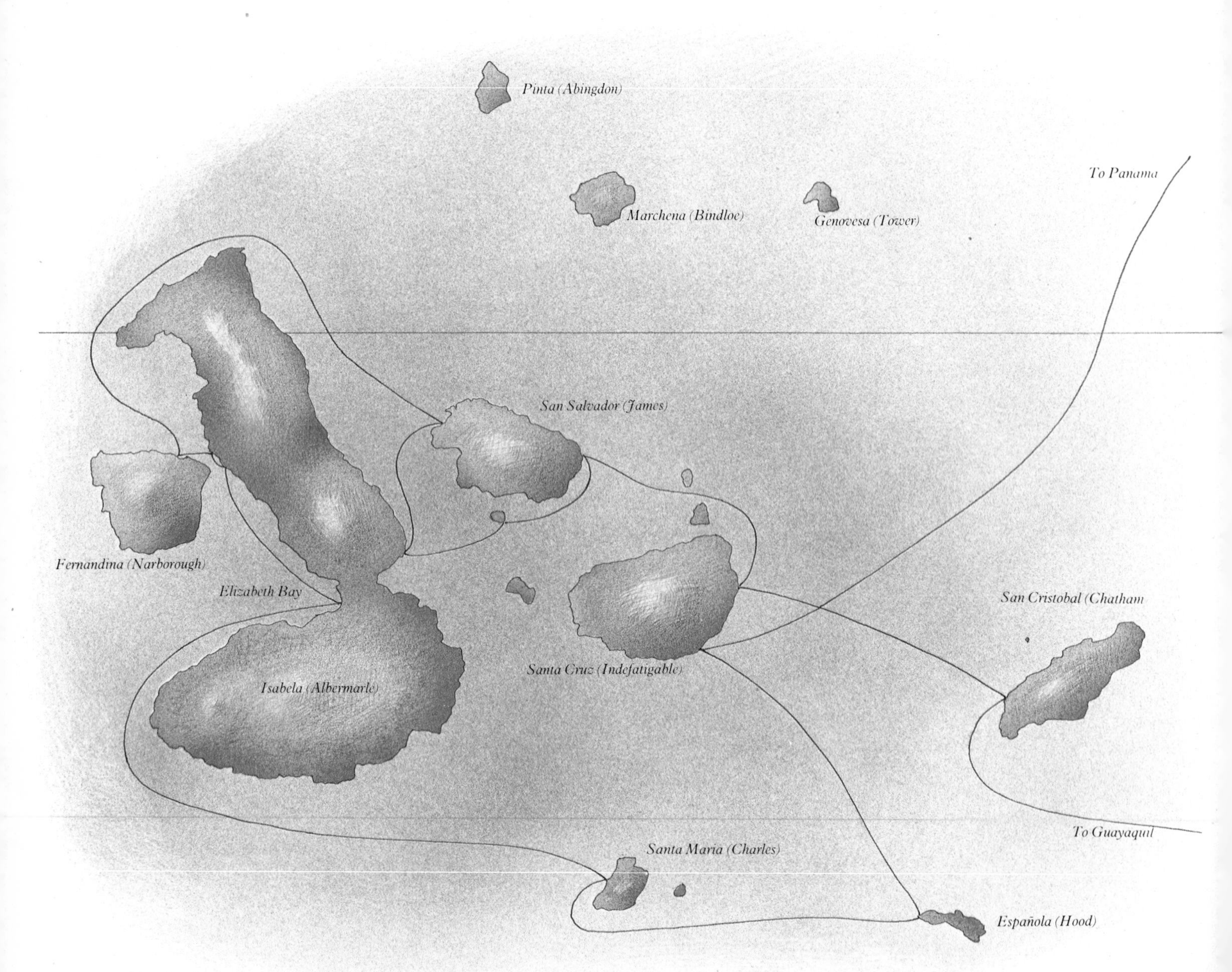

Charles Darwin's 'laboratory of evolution' (older English names for the islands are in brackets).

If any group of people with a deep interest in wildlife were asked to list all the places they would most like to visit, I would expect the Galapagos Islands to be in the top six for every one and the first choice for most. It is a unique place, as Charles Darwin found out in 1835, and still must have enough of the unknown about it to keep study groups busy until the crack of doom. 'A laboratory of evolution' is the description most often used and most befitting, and the Government of Ecuador, mindful of the islands' importance in other ways as well, is firm in its resolve that they should remain so.

The Spanish charts use the name 'Archipielago de Colon' and the romantics 'Islas Encantadas' (Enchanted or Bewitched Isles), yet it seems that 'Galapagos' has been adopted by all, this name being derived from the native giant tortoise (see page 136) – the most appropriate of all symbols.

The strange ambience of these islands – aspects of which, it must be said, tend towards the bizarre – began with the very nature of their discovery. This must be the only island group on earth to have been sighted first by a bishop – Fray Tomás de Berlanga was his name, Bishop of Panama, and he made his discovery on a voyage from Panama round to Lima, Peru.

The fact that the islands are some 600 miles off the direct course must not be interpreted as navigational shortcomings on the part of the clergy. It was a case of a flat calm and a strong westward set of the ocean currents. The good bishop was no more than a passenger on the whim of forces beyond even his control.

The Galapagos lie athwart the equator between 1°40′ north and 1°26′ south, and span longitudes 89°16′ to 92°1′ west. They are all volcanic in origin and number 61 officially named islands or islets, all with both Spanish and English names. Of these, 13 are of appreciable size; the total land area amounts to around 3,000 square miles, including a host of anonymous rocks and skerries. By far the biggest island in the group is Isabella (Albemarle), bigger in area than all the others together.

We humans are contrary animals. Often there is nothing more calculated to endear us to a place, or to a person for that matter, than vehement condemnation by another. Many a theatre critic must have noticed with ambivalent feelings the popular success of a production that was universally panned in the columns. Herman Melville, author of *Moby Dick* and other great works in the mid-1800s, fostered something less than total admiration for the Galapagos.

Here are examples from his *Encantadas*: 'Take five and twenty heaps of cinders, dump them here and there in an outside city lot; imagine some of them magnified into mountains, and the vacant lot the sea; and you will have a fit idea of the general aspect of the Encantadas, or Enchanted Isles. A group rather of extinct volcanoes, than of isles, looking much as the world at large might, after a penal conflagration.'

Frigatebird ♂ full of wind....

'In many places the coast is rock-bound, or more properly clinker-bound; tumbled masses of blackish or greenish stuff like the dross of an iron furnace, forming dark clefts and caves here and there, into which a ceaseless sea pours a fury of foam; overhanging them with a swirl of grey, haggard mist, amidst which sail screaming flights of unearthly birds heightening the dismal din.'

Charles Darwin, who must have shared a similar level of sensitivity to Melville, was at the very same time finding in these slag heaps the greatest inspiration zoology has ever known, and amassing notes for *On the Origin of Species* to be published 18 years later.

On a personal level I have to confess that I read Herman Melville's account long before good fortune and the *Lindblad Explorer* got me to Galapagos. Would he, I wonder, have found food for thought in his words being so widely taken as praise? Which ever way he had meant his writings to be interpreted, I for one saw them as a testimonial of no mean order and could not wait to see those flights of unearthly birds and listen to their dismal din. When I eventually did come to walk on that dross of an iron furnace, disappointment was noticeable only for its absence.

The Bishop had landed in 1535 and had written to his King with some very clear observations of the wildlife, including the tortoises, the iguanas and the tameness of the native animals; but he pre-empted Melville in his somewhat adverse comments about the terrain.

His was a turbulent time on the mainland, with Francisco Pisarro strutting about Peru murdering the Incas in the name of God and stealing their gold. The effects of the disquiet touched upon these islands in the shape of an escaped Spanish captain, apparently disgusted, on the run and anxious to avoid a retribution which could not have made enjoyable contemplation. He followed the path of the Bishop and found sanctuary on the Galapagos. He too told of the wildlife and the wonders. His name was Diego de Rivadeneira.

From then on, it was pirates of several nations; for a safe haven, for careening ships, for water and for tortoises. One of the great wonders is how these vast and vulnerable reptiles ever managed to survive. They were perfect provisioning – almost the forerunners of canned food. They could be stowed like crates in the hold. While life was in them their meat was fresh, and their life could be supported for weeks and weeks on end because that is the nature of the animal.

When the pirates' days were over the whaling fleets took on where they left off. Ships from New England bound for the Bering Sea round the Horn took in their supplies and called for more on their return.

On some of the smaller islands, the tortoises have gone for ever. Only the size of the larger islands, the heat and the hard walking must have preserved enough to get them over the critical watershed and maintain the population for today. Only here and on Aldabra – separated by two whole continents – can these great tortoises be found, yet their fossil remains are spread the world over.

Many factors contributed to the present fauna and flora of the islands: their distance from the mainland; the play of the Panama and the Humboldt currents; the nature of the landscape; the stocks of food in the ocean; and the *comparative* remoteness from any significant, large-scale human intrusion.

After the volcanic upheaval birds, as always, would have been the first colonists, accepting the newly-born and perhaps still-steaming land as a nest platform and no more. All their food came from the rich ocean and as they decorated the rocks with lavish offerings of guano in gratitude, they set in motion the beginnings of an environment in which plants could thrive. I often think of Surtsey, off Iceland, in this connection. When she was a ten-year-old island and just cooled off, she already had guillemots nesting on the firm basalt cliffs, terns on the cinders, and mosses already beginning to spread where the birds had sat.

A foothold of vegetation provides an environment for animals. Many of the Galapagos immigrants came from Central and South America, perhaps on rafts of vegetation that floated down swollen rivers and drifted across the sea. It must all have taken a very long time, but once here they stayed. Gales brought land birds miles off course in a manner which, as we have seen, can happen to a bishop. The new arrivals stayed because there was no need for them to do otherwise; but conditions were undeniably different from their homeland and so they began to adapt to their new and very unusual surroundings. Each species had lost the influence of its home territory, and only the Galapagos from now on would shape its destiny as well as its form. Some grew larger, some grew smaller, some birds that once flew became flightless through such plenty so close at hand. Adaptation of every kind began to meet a new set of needs.

Because some islands are well spread from others and vary in terrain, each island began to produce its own home-grown individual version from the same ancestral stock originally common to them all. All this Darwin saw, and saw it in perfect clarity: the concept that survival demands an ability to bend with the wind; that both the life style and physiology of animals are moulded by their surroundings; and that it is a process which, to use the vernacular, is ongoing.

Somehow in the Galapagos all this is condensed into a smaller space, making it much more apparent and easier to appraise. If one looks at the count of species, the very uniqueness of the wildlife on these islands is immediately clear.

There are some 57 species and sub-species of resident breeding birds.

Of these, about half could be described as either pelagic ocean birds or migratory species which one would expect to find breeding on other suitable islands elsewhere. The other half could fairly be described as 'land' birds, and of these no less than 25 are endemic – they are found nowhere else in the world. The plants are much the same. Of nearly 900 species, a quarter are endemic. Of the reptiles, practically every single one is either an endemic species or sub-species.

From the very beginning reptiles set the scene of these islands, and nothing else leaves so lasting a memory. The evening sun on a beachful of iguanas, statuesque against the sky, is something straight from prehistory. Herman Melville, in *The Encantadas,* again had things to say about this species and the general preponderance of reptiles:

Vermilion Flycatcher
"Brujo"
Pyrocephalus rubinus.

'Another feature of these isles is their emphatic uninhabitableness . . . Little but reptile life is here found; tortoises, lizards, immense spiders, snakes and the strangest anomaly of outlandish nature, the aguano (iguana).'

Of the islands themselves he went on to say: 'No voice, no low, no howl is heard; the chief sound of life here is a hiss.' *The Encantadas,* 1854.

Reading between the lines, one gets the impression that Melville may well have been the one man in history whose sojourn in the Galapagos Islands actually fell somewhere short of enchantment.

Marine iguanas sample the sea air at Punta Espinosa, Fernandina Island. More than any other animal, these lizards contribute to the strange air of prehistory that lurks in the Galapagos. They lie about in little groups and multitudes, like mini-dinosaurs, timeless and unmoving. Diving to 30 feet or more, they stay submerged for an hour as they graze on the green algae that cover the undersea rocks.

Like so many Galapagos animals there are local races of marine iguanas on different islands. Seven variations in size and colour have been recorded, with the largest specimens on Isabela.

OVERLEAF Rainbow over Isabela. Every island has an English as well as a Spanish name. The English version of Isabela is Albemarle, the biggest island of the group. The Spanish name for the entire archipelago is 'Islas Encantadas' – the Enchanted Isles, a name that no one would dispute.

ABOVE Sealions at Turtle Cove on the island of Santa Cruz. Galapagos sealions are closely related to the Californian sealion of the mainland, which was certainly their ancestor. The cool Humboldt Current bringing waters north from the sub-Antarctic carries with it rich stocks of fish. So the sealions have stayed here in a land of plenty, athwart the Equator.

OVERLEAF Spot the mammal . . . a Galapagos sealion's proud pose, silhouetted in a wave-made window on James Island.

LEFT Mustard rays drift by slowly like giant autumn leaves, 'flying' with the most leisurely of wingbeats. A shoal could be 50 strong and stretch from the surface to an oblivion of depth.

ABOVE Turtle Cove on Santa Cruz Island was named for the most obvious of reasons. Pacific green turtles are mating here. Before mounting, the male often bites the skin behind the female's head – leaving pale, calloused scars on her shoulders and neck.

TOP Like a painting-by-numbers crustacean, this Sally Lightfoot crab peers into the grounded camera. Crabs abound, scavenging the rocks and beaches like so many ten-legged vacuum cleaners.

ABOVE The Galapagos snake – Punta Suarez, Hood Island. Once more the magic figure seven: seven sub-species of this non-poisonous snake are scattered through the islands. They grow to about three feet long and some specimens are beautifully marked.

RIGHT A Galapagos-trained herpetologist would know instantly that this lava lizard hails from Santa Cruz Island. The greyish general colour, the rusty patches and the patterning of dark spots give it away.

Its scientific name is *Testudo elephantopus* – in Spanish, 'galapagos', the giant tortoise that gave the islands their name. Here, and Aldabra in the Indian Ocean, are the only sanctuaries for a species that once roamed the grassy plains of every continent.

Lava fields at Sullivan Bay, James Island. Humans seem out of step with such terrain. Their feet never find a smooth or level spot, while lizards scamper and snakes slither with irritating ease.

ABOVE Giant cacti carpet some areas of the islands. These prickly customers are on Santa Fé.
RIGHT As photogenic as you please. Galapagos sealions pose at Plaza.

A ghost crab casts a ghost crab's evening shadow across Espumillia Beach, James Island. In a variety of colours, all pale, ghost crabs are the most nimble running crabs of all. They have been clocked at over five feet per second – like the rest of their fellows, sideways.

Galapagos and Beyond

Galapagos Penguins – Fernandina

'With another pre-sunrise start, we walked to the highest point on the cinder cone of Bartolomé Island. From this vantage point we could see the lava floes of James Island, as well as the islands of Santa Cruz and Isabela (15 and 25 miles away). It also gave us the 'classic view' of the Galapagos – the isthmus of Bartolomé and Pinnacle Rock, with James Island in the distance.' From Tom Ritchie, 27th April 1978.

There cannot be many lucky enough to visit the islands, who have not stood on this spot. It once afforded me the pleasure of walking up there with Eric Shipton, the mountaineer, and to be in such a place with so lovely a man is not easily forgotten. He worked in the ship often then, as an expert on several aspects of the Galapagos, and many people profited from his knowledge as well as his company.

Normally the ship came across from Guayaquil, having picked up her passengers flying down from Quito, but then she would stay in the islands. Passengers flew out from the air strip at Baltra in the same chartered plane that brought in the new ones. The final voyage might end in Panama or farther up the coast because the *Lindblad Explorer's* visits there tended to be on her general northward trek from March into May, towards a season in the Bering Sea for the summer. She has not visited for some years now since the tariffs that one encounters in port were so increased for ships not under the Ecuadorian flag that this made it impossible to break even.

The voyage across was generally placid and the sea very blue, making a nice clear background for the curious spouts of sperm whales which were regularly recorded. Few animals are asymmetrical, and such exceptions are always striking: the sperm whale's blowhole is very far forward on its great square head, and emits the spout not only slightly ahead but significantly to port.

We always called at San Cristobal, the nearest of the islands to the mainland and also the capital, and went on to Academy Bay on Santa Cruz, the home of the Charles Darwin Research Station. This famous station grew from the Charles Darwin Foundation, an international body, and was established in 1968 with Raymond Lévèque as its first director. The Foundation has done much to achieve the status of a national park for the Galapagos and a binding conservation message throughout the islands.

It is interesting how many of the *Lindblad Explorer's* scientific staff worked on research at the station before joining the ship: Michael Harris, who wrote the *Field Guide to the Birds of the Galapagos*; Soames Summerhayes and Jim Snyder, both of whom have been expedition

leaders; while Roger Perry, Eric Shipton, Roger Tory Peterson and Michael Castro were constantly in and out of the station before being in and out of the ship. 'Rembrandt of the Shutter' Eric Hosking was here too as a ship's naturalist, but he was rash enough to admit in public that he repaired his own cameras and thus it was that his cabin became a photographic workshop. In addition he was faced with the unenviable task of breaking it to distraught passengers who had just recovered a valuable camera from the ocean, that they might as well have left it there!

But this is very much Jim Snyder's own country. From his log, 29th March 1978: 'Islas Isabela and Fernandina . . . Sealions and flightless cormorants, pelicans, lava herons, Sally Lightfoot crabs and Galapagos hawks crowded the narrow peninsula of lava and *Brachycereus* cacti, and we marvelled at the thousands of dragons (marine iguanas) herded together'.

Later he had things to say about his favourite of all birds: 'And then there were blue-footed boobys . . . As we wandered among these comical sulids, their blueness varied with each from pale turquoise to deep ultramarine. They paraded around with great agility, proudly flashing their bright webs with high exaggerated goose-steps. A breeding pair promenaded about, their feet raised slowly with ridiculous deliberation, the male whistled, the female honked, while they rattled their heads in unison and gently touched bills . . . And so with bowing, wing-presenting and mutual sky-pointing, the pair whistled and honked themselves together – the start of a booby love affair.'

In time Jim acquired skill in imitating the boobys he loved so much and was apt to cause embarrassment to his shipmates by embarking on the performance in a public place. When he got to the sky- pointing, wing-presenting and goose-stepping bit, he gave it his all and fairly trembled with ecstasy. Onlookers, who tended to be unfamiliar with the choreography of sulid courtship, or indeed the purpose of the exercise at all, betrayed through their expressions certain doubts about the fellow's sanity. I think we all secretly wondered if he ever went through the routine in private or tried it out on mates of his own choice.

'30th March. Long before sunrise a little band of mountaineers set out up the slopes of Alcedo volcano (Isabela Island), their quest, the giant tortoises and an ascent into the heart of Galapagos. . .'

Certainly these were the biggest tortoises we ever saw and Isabela supports the largest total population in the islands, something over 6,000. The big ones were high on the upper slope of Alcedo, often advertising their presence by the grunting and heavy thuds of rival males, weighing up to 500 pounds each, charging one another in slow motion and heaving themselves into long and ponderous trials of strength.

There are 11 surviving races of tortoise on the Galapagos out of an original 14, and five of these are found on Isabela where each great volcanic cone has its own population. The Alcedo volcano is one of six on this island but not the highest. Its even crater rim stands at 3,700 feet and within it lies a sunken flat caldera about two miles across. From the summit the view is truly gorgeous. There are plumes of steam, for five of the volcanoes are active, and lava fields cover a good half of Isabela's 1,800 square miles of surface.

To the north-west, past the cone of Volcano Darwin, can be seen Volcano Wolf, the highest point in the islands at 5,500 feet; to the south, across the Perry Isthmus, are the wide-topped crater and ridges of Cerro Azul and Sierra Negra; across the generally eastward sector, the quiescent cones of the other islands stand from the sea. It is easy to be carried away by distance until the sudden beacon-red spark of a vermilion flycatcher brings the attention to heel. It is a good place for these birds; I have often noticed how a breeze over a sharp ridge will bring insects circling into the eddy in its lee. With luck a Galapagos hawk may come by, using the windward side for a spell of free lift and an effortless transit of the mountain.

I am sure everybody finds a pet place of their own. I have mentioned two out of a hundred possibles, but there is one more that in a perverse sort of way appeals to me enormously, perhaps because Herman Melville would have found it so disagreeable. It is the Devil's Crown.

The 'Crown' is all that remains of a broken-down crater core rising from the sea off the island of Santa Maria. Between the broken teeth the water is turquoise, filled with coral, and the coral quite naturally is spiky. Outside the teeth there is a drop-off and a strong set of current, and here row upon row of hammerhead sharks lie head to tail right down into the depths, stemming the flow with languid wafts of the tail.

But it is the teeth themselves that are so unremittingly hostile. They are grey and look like gigantic sculptures, carved in the overall likeness of a frayed sponge and cast for posterity in a confection of concrete, razor blades and broken glass. As if this were not hostility enough, each tooth is surmounted by a harpoon-bearing cactus and suchlike adornments of botanical barbed wire. The Devil's Crown was aptly named; one can all but feel the prod of his trident.

Then across the scene comes a tropic bird, smooth and white and soft, its form a perfect harmony of discretely curving lines, its flight the poetry of motion itself, and no more exciting contrast in character could have been struck.

It was the bird life of the Galapagos that catalysed Charles Darwin's ideas concerning the origin of species. In particular he was struck by the Galapagos finches, 13 species of rather drab little birds that nevertheless had a pivotal role in revolutionising the biological sciences. From his notes on the visit:

'The most curious fact is the perfect gradation in the size of the beaks in the different species, from one as large as that of a hawfinch to that of a chaffinch. Seeing this gradation and diversity of structure in one small, intimately related group of birds, one might really fancy that from an original paucity of birds in this archipelago one species has been taken and modified for different ends.'

Besides such 'fanciful speculation' that was to evolve into a cornerstone of modern thought, Darwin also sprinkled his writings with descriptive passages. Of the giant tortoises: 'Near the springs it was a curious spectacle to behold many of these huge creatures, one set eagerly travelling onwards with outstretched necks, and another set returning, after having drunk their fill. When the tortoise arrives at the spring, quite regardless of any spectator, he buries his head in the water above his eyes, and greedily swallows great mouthfuls, at the rate of ten in a minute.'

Jim's notes for the Galapagos end with idle days northward bound. '6th May – at sea with dolphins and pilot whales. 7th May – at sea on a glass-smooth pacified Pacific. 8th May – Acapulco.'

Blue-footed boobies at Tagus Cove on Isabela. These are very communal birds, feed close inshore, and in consequence are the most frequently seen of the three Galapagos booby species. They plunge like javelins into schools of fish but do not take their prey on the way down – instead they seize it against the light as they rise to the surface.

ABOVE The pair-bonding ceremonies of the waved albatross include enchanting sequences of this gentle 'bill-fencing'. The tiny island of Hood, in the extreme south east of the Galapagos, is not the only breeding ground for this, the largest seabird of the eastern Pacific, with a total population of about 10,000 pairs. A few pairs nest on Isla La Plata near the Ecuadorean mainland.

RIGHT One family from a cliff-nesting colony of swallow-tailed gulls on Tower Island. Galapagos is synonymous with the unusual, if not the unique. So it comes as no surprise to find that this gull, with its large, dark eyes, is the only gull that feeds by night and is the only gull that habitually lays a single egg.

LEFT Portrait of a masked booby, Tower Island. The masked, the largest of the boobies, is more of an ocean wanderer than its cousins. Certainly the adult plumage of sharp black and white also makes it the smartest. Birds like these would have been among the first settlers after the great volcanic upheavals, some three million years ago, that raised the Galapagos Islands out of the sea.

ABOVE 'Give me one more feather' screamed Dennis Puleston's nearly naked parrot, 'and I could fly like a *** eagle!' The Galapagos flightless cormorant, with about 800 pairs, is one of the world's rarest birds.

FAR LEFT On Tower Island a male greater frigatebird or man'o'war bird limbers up, his inflated throat sac like a splendid red spinnaker well set in a brisk force five and his iridescent green wings shivering at any female flying by.

ABOVE And here she is, seduced by one of the most exhibitionist courtships in the avian world. On the other hand . . .

LEFT . . . failure to impress a female produces in the male a sorrowful reaction of both physical and mental deflation.

THE ATLANTIC ARCTIC

A selection of far-north routes, including the triumphant North-West Passage.

There is nothing like a good school geography book to put things simply enough to last for life.

The difference, the primers all say, between the Arctic and the Antarctic is that while the Antarctic is a continent surrounded by ocean, the Arctic is an ocean surrounded by continents. Although in area the winter sea ice of the Arctic ocean just about matches the Antarctic land mass, that is where the comparison ends. The fact that submarines have actually crossed the North Pole says it all. The abyssal plain in that area is around 2,300 fathoms.

There are other differences, too. The Arctic can boast a long list of properties the southern continent never knew. The Arctic has indigenous people, several predatory land mammals other than the people, and herbivorous land mammals that are preyed upon by the predators including the people. It has beautiful wild flowers and a host of highly sophisticated insects – some would say too sophisticated and, in places, a great deal too numerous for comfort. ('Shall we eat him here or take him away and have him at leisure?' is what one Alaskan mosquito was reported to have asked another.) You may die of hypothermia with ease in Antarctica, but you will not be plagued by blackflies, midges or mosquitoes, and that makes a big difference to the quality of life.

Arctic birds are varied and plentiful. From lands throughout the northern hemisphere, and even from south of the Equator, migrant species head generally north in spring, some reaching to the farthest limits of tundra and scree to nest. The result is that the check list of breeding birds north of the Arctic Circle is immense compared with the same latitude south. It includes ducks and geese, waders, small land birds, seabirds of all kinds – gulls, skuas, terns, auks, guillemots and puffins – and even handsome avian predators like the white falcons and the snowy owls.

There are even trees in the high Arctic, forest trees – but they grow only to ankle height, and although they turn beautiful colours of red and gold in autumn and shed their leaves, these fall only a few inches to the ground. Some of the trees are immensely old by standards we might use in Britain or the United States for some gnarled historic village oak. Conversely by Arctic standards the equivalents of such oaks would be vast and spreading – growing at least to knee height!

Earlier I mentioned that in 1972 the *Lindblad Explorer* had gone farther north than any passenger vessel had ever been, to 82°12′. It was in fact Hasse Nilssen's first voyage as captain, and Lars-Eric Lindblad was on board too. Sadly we were tied to a schedule then and had no time to press on even farther; the ice was loose, the weather warm and had time been on our side we might have almost reached the North Pole – or so it seemed at the time.

As it was the voyage had many other high spots, including a circumnavigation of Milne Land, which might or might not have been possible had we lingered in the ice. Milne is a vast island of east Greenland at the back of Scoresbysund, and from the charts the channel that separates it from the mainland looks irresistibly inviting, so we decided to try it.

Now, Scoresbysund is not the sort of 'port' that provides official pilots, but local knowledge was to be had along with a great welcome and willingness to help. Jacob, a much revered Eskimo hunter, said he would show the way because he had done it often. So with Jacob on the bridge

at the Captain's elbow, we sailed confidently into the channel.

The cruise provided everything one could have wished for in both wildlife and landscape, and took most of the day to complete. As we approached our starting point in the wide water of the sound, Hasse Nilssen casually asked Jacob what other ships he had piloted through. Jacob smiled a huge smile and with his arms made a gesture of paddling. 'A kayak, just a kayak' he said. The *Lindblad Explorer*, we thus discovered, was the first 'ship' to try it, and is probably the only one.

You win some and you lose others, as philosophers say; and on that voyage we 'lost' a passenger. I believe it was the only time that this has actually happened, though on several occasions it has come close.

There was a thoroughly sound system with numbered tags for checking all aboard, but through a convoluted chain of circumstance I now forget, the system suffered a hiccup. We sailed away in dense mist from Svensköya, an island of Spitzbergen, with one of our passengers unknowingly still ashore and sharing the island with five polar bears. I will not dwell on the drama; happily we got this splendid lady back, clutching her plastic bag of botanical specimens and having suffered nothing more than a fairly long period of unease. She was not, in that respect at least, alone.

Lindblad Explorer has operated in the Arctic each year in the boreal summer but almost exactly half the circum-polar possibilities have been denied her – the area from the Barents to the Chukchi Sea, the Russian Arctic. This north-east passage had been first achieved by the Swedish baron, Nordenskjöld, whose nephew Otto's exploits are mentioned in the Antarctic context. The year was 1878-79, and his ship the *Vega*. When the centenary came up in 1978 it was planned to try to take *Lindblad Explorer* through the same route, but sadly no clearance could be obtained. Had it been, we might have made it – or perhaps, like the baron himself, been stuck there for the winter. Her Arctic activities therefore have been tied to northern Scandinavia, Bear Island and Spitzbergen, Jan Mayen and Iceland; then from Greenland westwards through the whole Canadian Arctic to Alaska. It is all very varied and (in so far as sour grapes have their uses) much more interesting, more scenic, and much more workable for a ship than the Asian sector.

Up here the Atlantic Ocean plays strange games with the climate. The Gulf Stream gliding up from Florida crosses the ocean to lay tropical jetsam on the west-facing beaches of the British Isles. On up north it travels to the tip of Scandinavia, spreading a benign climate where tall trees can grow, girls can wear bikinis and the temperature can reach the 80s F in places like Tromso and Hammerfest – nearly 300 miles north of the Arctic Circle.

Hammerfest is something like 70° north; at the same latitude across the ocean is Scoresbysund, Greenland, with Jacob and his kayak, ice in the fiords and the temperature averaging about 16°C (30°F) lower. It is often said that Greenland and Iceland should trade names. Iceland is touched by the Gulf Stream too and all of it is below the polar circle. In Greenland terms it is a nice warm place, and not just because of the bountiful intrusion of active volcanoes. The Denmark Strait separating the two is the basin of the east Greenland current, a desperately cold south-flowing stream that garners lofty and spectacular icebergs, carved off the glaciers which thrust down from the Greenland ice cap, and propels them remorselessly into the shipping lanes of the Saint Lawrence and even farther.

It was in these waters just off Cape Farewell that we had the best sighting ever of blue whales. There has always been a whale watch maintained in *Lindblad Explorer*, and in this way sightings with their coordinates can be faithfully recorded.

'4th September (1972). On course Kulusuk to Cape Farewell, 1630 hours. A large pod of rorquals sighted on our starboard quarter, travelling very fast and on parallel course. They overtook us easily and we soon had them close abeam and slowly drawing ahead. They were easily identified as blues. As far as we could see there were about 15 with a school of pilot whales among them, the blues spouting almost in unison and to the height of the bridge wing . . .'

They, or others, appeared again the following day – this time without the pilot whales. I believe there has never been a more impressive viewing from this ship of the largest animals that have ever lived on earth. The statistics from the lectures kept running in my head – a full-grown blue whale equals the weight of 25 African elephants or 2,000 average human beings.

Strangely enough, I do not think we have ever recorded a blue whale south of the equator in all those 15 years of cruising. When I was a little boy on the way to live in Australia with my parents and brother, we saw blue whales every day on the leg from Capetown to Albany. We saw other species, too, all of which were identified for us by the ship's doctor who had worked in South Georgia. But children fall victims to their own nimble imaginations and I remember being bitterly disappointed at the first one, after the build-up my mind had been offering. A big fish, yes, I thought discouragingly, but one helluva lot smaller than the ship!

That was a deflating voyage, too, over my first albatross, which I had expected to be about level pegging with a Boeing 747. Even when one came aboard through a minor flying mishap and clapped its bill at me, and I was able to touch it, I still felt badly let down. Just a big, clumsy seagull, I thought. Fortunately, the process of readjustment is swift at that age.

This pile of notes from the Arctic voyages is full of half-forgotten things, and make the most nostalgic reading. One entry tells of a landing late one night to walk on the moraine of Barentsøya, Spitzbergen: 'Reindeer, surprisingly tame, with cast antlers all over the landscape. We then came across the enormous half-buried subfossil jawbone of a whale, too big for anything but a blue. It lay flat and had formed a natural dam across a melt-water stream, creating a deep clear pool above it, six yards across. There were other whale bones all over the slope, and back at the ship we measured the height – 150 feet above sea level.' The whales would have been natural strandings, left high and dry as the last Ice Age came to its end.

There were accounts of Amsterdamøya and the remains of the eighteenth century Dutch whaling station, when migrating right whales were so plentiful that they could be harpooned from the beach. And the little island of Kadavaøya, still littered with human bones from the plague that struck the whalers.

On Danskøya the collapsed hut, the bunkers of rusted iron shavings and the Doulton glazed earthenware carboys for hydrochloric acid, like a giant's chemistry set for making hydrogen, are all that survive of Salomon Andrée's ill-fated attempt to cross the North Pole in a free balloon in 1897. But so often it is the trivial little note that says the most about the Arctic: '. . . crowds of snow buntings came aboard (*Iquatck*) and hopped about till dusk . . .'

'Tonight, 3rd September, a bright new display of the northern lights crossed the whole sky . . .'

With this last mention came a recollection of my mother (she of the ninth beatitude). She was leaning over the rail watching a water spout as it writhed along below its sinister funnel-shaped cloud. 'There is nothing like a natural phenomenon' she said, 'worth all the man-made ruins in Christendom – and most of those outside.' Water spouts, waterfalls, geysers, volcanoes, great canyons, good thunderstorms with torrential rain, the break-up of ice on the Yukon, all filled her with the deepest wonderment, but her favourite of all was the Aurora Borealis.

ABOVE '. . . the bear and the Zodiac boat played hide-and-seek all morning.' The reward of patience is this coy portrait on Moffen Island, Spitzbergen (Svalbard).

OVERLEAF Polar pack ice and passenger, almost anywhere. A picture that says everything about bears and their nomadic summer lives – an unhurried seal-hunt over a patchwork of water and lily-pads of ice, with the horizon their only limit. Theirs is a circum-polar quest, linking the Western World with the Soviet Union between the high 80s north and the Arctic Circle. It is a long day for the ice bears; there will be no night until autumn.

BELOW Parental care fulmar-style on Bear Island, between the northernmost tip of Europe and Spitzbergen. The fulmar is one of those thrusting offshore birds that is actually extending its breeding range. A northern, vest-pocket version of the albatross, it likewise lays a single egg. The prominent tubed nostrils are part of the bird's built-in desalination plant.

RIGHT The curious volcanic sea-cliffs of Heimaey, off Iceland – a picture that calls for a sound-track. In full cry, a kittiwake colony that has the incessant background din of a children's playground.

TOP The walrus is one of those very select animals that is the only species in its genus. The genus is *Odobenus*, the species *Odobenus rosmarus*. There are said to be two races, an Atlantic and a Pacific. These individuals, from Ellesmere Island in the Canadian high Arctic, present an unanswerable question of identity.

ABOVE Loungers of the Atlantic race, *O. rosmarus rosmarus*, near Spitzbergen. This race, distributed from the Kara Sea to west Greenland, provided the few animals that once visited northern Britain.

A wave-cloud over the peaks of Hornsund, Spitzbergen.

The yearly climax of a life of travel – two eggs in the nest scrape of an Arctic tern at Ny Alesund, Spitzbergen. No creature experiences more sunlight than this global traveller. Each year it commutes a minimum of 16,000 miles to follow summer's transit across the world.

LEFT A view of Cape Searle, Qaqaluit Island, in the Davis Strait. There is a certain character and shape common to both rock and ice alike, making colour the only difference.

ABOVE These giant icebergs float in Disko Bay, Greenland. The enormous flat-topped, tabular bergs of Antarctica are unknown in the north – there are no ice-shelves to produce them. Instead intricate and bizarre castellated shapes, calved from glaciers, take their place. Perhaps the best area hereabouts, for icebergs of real personality, class and size, is Melville Bay in north-west Greenland. Many still show the rugged cleavage lines where they broke free, although the lower levels have been eroded and reshaped by waves into smooth, sensuous curves. These are of the kind that by drifting south into busy shipping lanes could sink a *Titanic*.

OVERLEAF Classic Arctic icebergs often retain the rugged, rock-like character of the glaciers that bore them to the sea, long after wave action has moulded their lower levels into simpler, more graceful lines.

BELOW LEFT 'Rotten ice' of the summer thaw, off Cape Seddon, Baffin Bay. Beyond is the edge of the Greenland ice-cap. This coast is an age-old hunting ground of the Inuit. The first Europeans to explore it were William Baffin himself and Robert Bylot in 1616.

BELOW Dovekies (little auks) at Hartstene Bay, Greenland. These plump little birds are about the size of a lemon. They fly like bumblebees, swim and dive like sea-ducks, and are the smallest of the auk family. They breed in crevices of screes and rockfalls high up into the Arctic. By late summer Baffin Bay is full of them, all moving generally southward before the sea begins to freeze over. They spend the winter on the open ocean.

The rocky buttresses of Cambridge Point, Coburg Island. A look of peace and tranquillity belies the closer reality of noise and bustle and multitudes beyond belief. Throughout the summer months, hundreds of thousands of thick-billed murres and kittiwakes congregate here.

North to Alaska

Bubbling up from my schoolroom subconscious like methane from the bottom of a pond, comes a fact about Alaska which I find a lot more interesting than some of the other little bubbles that reach the surface. Alaska, with all its outlying islands of the Aleutian chain, was sold by the Tsar of Russia to the United States in 1867 for $7.2 million – the lot.

Congress at the time was appalled at such robbery, and I have forgotten how few cents per acre that figure represented, but it was something like two. In those days Alaska was a 'Territory'; today it is a State of the Union, signed by President Eisenhower in 1958, and one cannot help pondering on the degree of chagrin that must have prevailed in some areas over a brand new state that was actually bigger than Texas – in the interests of exactitude, two and a half times bigger.

Everything about Alaska is on the same grand scale. If you measure the distance between the south-eastern and south-western extremities of the state, from somewhere near Rudyerd Bay to the Aleutian island of Attu, it is equal to the entire span of the United States.

In Mount McKinley, over 20,000 feet, Alaska has the highest mountain on the North American continent. The mighty Yukon, rising in the Canadian territory that bears it name, is North America's third longest river and among the world's top dozen. Its basin divides the state almost exactly in two, separating the rounded, glacier-free mountains of the Brooks Range from the really high ground of the Alaska Range – Mount McKinley and the Saint Elias-Mount Fairweather massif – that forms the coastal mountains to the south.

Even the largest island under the American flag, Kodiak, is in Alaska, supporting the world's largest race of the brown bear.

With such a span and with such terrain, the climate has to be varied. Basically it is what is euphemistically described as 'continental', a system that ensures warm, sometimes humid, summers interposed with winters offering what can only be described as brass monkey weather.

I have offered up this potted geographical profile to try to set this utterly gorgeous place in some sort of frame. It helps, I feel, to know what is inland even if we are seeing little more than the coast, beyond perhaps a glimpse of the cloud-wrapped summit of McKinley on the local flight from Juneau up to Nome. But for all that, a ship sees much of Alaska.

Somewhere a statistic must exist for the total length of her coastline and how near it would reach, if straightened out, to the moon. But the greatest thing about Alaska, starting in the misty, rainy, cloud-shrouded pan-handle of the south-east, is that there are waterways of pure ocean, fiords deep and clear and calm. There are hemlock and spruce trees to the land's edge, but one is constantly reminded that the flat water in which our ship rests is ocean, salt water, and fiercely tidal. Buoys marking channels often have hissing white bow waves and on occasion you will see one hove down, to shear about in protest as it disappears into the deep. Here and there new rocks appear, even as you watch, as though some gargantuan bath had pulled its plug only to refill again later at the same remorseless speed.

And up here came Captain Cook. As ever, he managed well. He must have played the tides with the greatest skill. For the wholly romantic amongst us, to anchor in one of Cook's havens is evocative in itself. Christmas Bay, Kerguelen, Possession Bay, South Georgia, the Cook Islands to the east of Samoa with the waving palms, then up to Cook Inlet, Alaska, and the lovely, misty, spruce-bordered waterways up to Anchorage itself.

Heading north by sea the pan-handle is a logical beginning. There are places here where a ship may come in close enough to the shore, in safety, to touch the trees with her stemhead. There is a land-locked, lake-like atmosphere with the loamy smell of the forests. A cathedral hush falls on people in places like this, for fear of disturbing a bear that is nonchalantly combing the beach a few yards away.

Bald eagles are there at every turn of the channel. Black curtains of rock tower hundreds of feet high, burnished like marble by millenia of grinding ice and laced with cascades that may splash down on to the very fo'c'sle from vertically overhead.

Someone swats a mosquito on the back of the hand. Other people say 'Shhh'; the bear might just look up, or he might shuffle off into the spruces to keep a none-too-pressing appointment elsewhere. There could be a mountain goat up there against the sky, looking quizzically down, the white coat haloed with sunlight. There might be porpoises under the bow and Steller's sealions within a stone's throw on the rocks. Or an exhibitionist humpbacked whale, swimming upside down, may roll over languidly to thrash the water with 15-feet flippers before breaching clear out of the sea in one tremendous slow-motion upheaval, as if he were a reincarnated salmon.

Glacier Bay. The ship is suddenly and violently rocked by a massive calving of ice that prompts anxiety for a bottle standing on the cabin table below. Bald eagles balancing on flailing wings use their talons as crampons to grip the bobbing, slippery peak of an iceberg newly born and floating free.

'19th July, Middleton Island. At the head of the beach lay the disintegrating wreck of the USS *Coldbrook*, a military supply vessel torpedoed in World War II, now completely taken over by kittiwakes. Every suitable spot on the vessel, including the crow's nest and anchor flukes, was occupied by a nest.'

Middleton lies in the Gulf of Alaska, south of Prince William Sound, an island of birds and flowers. The wreck of the *Coldbrook* is visible for miles, a great rusty landmark, and the cries of the kittiwakes amplify the eerie acoustics of her empty iron shell. The cliffs above are lined with puffins and murres; the strip between, a mass of white orchids.

To me this proliferation of wild flowers is the most lasting single feature of Alaska. The beaches, the inland bogs, the tundra, forests and the mountain slopes are all the same, more flowers than grass. Flowers so profuse they inhibit walking for fear of treading them underfoot.

The villages and little towns of the Aleutians are geared to fishing and to fish processing. Still essentially Russian in appearance, the church architecture is inspired by the onion rather than the spire and filled with icons and traditional trappings of Russian orthodoxy. The same is true of the Pribilofs, Saint Paul and Saint George, in splendid isolation in the shallow Bering Sea. Northern fur seals are culled here annually in large numbers under licence and the controversy it promotes is every bit as vehement as the issue of hunting the fox in England. There are arguments of sound economics and preservation of livelihoods, and there are counter-arguments concerning the whole ethical spectrum of wildlife exploitation in whatever form. But to me, one of the tragic byproducts of an Arctic where hunting was always a respected and prevailing tradition, and time tends to hang heavily in the long hours of summer daylight, is the brutalising effect of present-day slaughter.

All that moves is a target, whether it is recoverable or not, whether it is edible or not. Trigger-happiness becomes a conditioned reflex of human existence and nobody seems prepared even to shy a question at its wisdom. Around the settlements a major and ever-increasing constituent of the topsoil is empty cartridge cases, while the creative product of the heart and mind seems to dwindle. To an outsider, even to one brought up under very parallel persuasions, I find it saddening.

'18th June . . . As we cruised around the rocky shore line (Round Island, Bristol Bay) we could see Steller's sealion and walrus massed on the beaches; thousands of murres, kittiwakes and puffins, hundreds of parakeet auklets were nesting and roosting on every available ledge. . .'

The walrus spectacle of Round Island is something one could never forget. Those great torpid heaps of all shapes, hundreds of them, tusks pointing in every direction smooth and glistening against the roughness of their hides. Once there must have been a thousand places where this spectacle could be seen – almost within living memory there were walrus in the Outer Hebrides – but Round Island now is almost alone.

I was happy to be there for a whole day while Franz Lazi, the West German film-maker, and Des and Jen Bartlett filming for Survival Anglia, were working on the crowd scene from the cliffs above. The cameras with huge lenses picked out the by-play, panned over the throng and rested, the cameramen watched with patience of their kind. Cameras to me are another of life's deep mysteries, but I love to be with professionals when they are working, and not just for their good company. There is no greater satisfaction than watching a very special happening and knowing that it will be there, recorded for ever and for others.

Meanwhile, '. . . some of us wandered off to enjoy the splendid display of wild flowers now at their peak, and to watch the constant traffic of seabirds between the sea and the cliffs, but by now it was blowing very hard, so we took another wet Zodiac ride back to the ship and were soon on our way north.'

North meant a sight of musk oxen on Nunivak Island, moving over the tundra like Lascaux cave drawings, shy and nimble; and Inuit fishing camps with drying salmon, taken from a wide, shallow river of elusive channels and firm sandbars. There were emperor geese, too, coming into pools behind the coastal range of sand dunes, a wide and lovely open landscape with a great deal of sky.

Saint Matthew Island. A dream goal of many a 'life-lister' or 'twitcher', those indomitable birders who keep a check of every species they see. Saint Matthew has its Asiatic vagrants, sometimes in abundance; here the little white McKay's snow bunting abounds, and I never cease to derive a welcome reassurance from finding that what the book says, is just the way it is.

To Diomede Islands. The Polar Circle cuts north of the Bering Straits and crosses Alaska just south of the Brooks Range. So the Diomedes, in the narrowest part of the Strait itself, are not truly Arctic, but the thing that makes them so special is the International Date Line.

Little Diomede, the eastern one, is United States; Big Diomede, just two and a half miles to the west, is Soviet Russia; and only the massed kittiwakes may share both places without let or hindrance.

I got the feeling of a vast telescope watching me on the only day when both islands were infinitely clear. I climbed to the top of Little Diomede and from there the whole of the Seward Peninsula of Alaska stretched away into oblivion behind me. To the west lay the narrow strait with the tiny *Lindblad Explorer* below and Big Diomede, USSR, rugged, apparently untouched and even through binoculars maintaining its disguise of the uninhabited. Over the top of Big Diomede could be seen, in the far distance, the most easterly cape of mainland Asia, Cape Dezhneva, lying across the horizon. And it came to me strangely that while I was immensely enjoying today, I was actually looking at tomorrow.

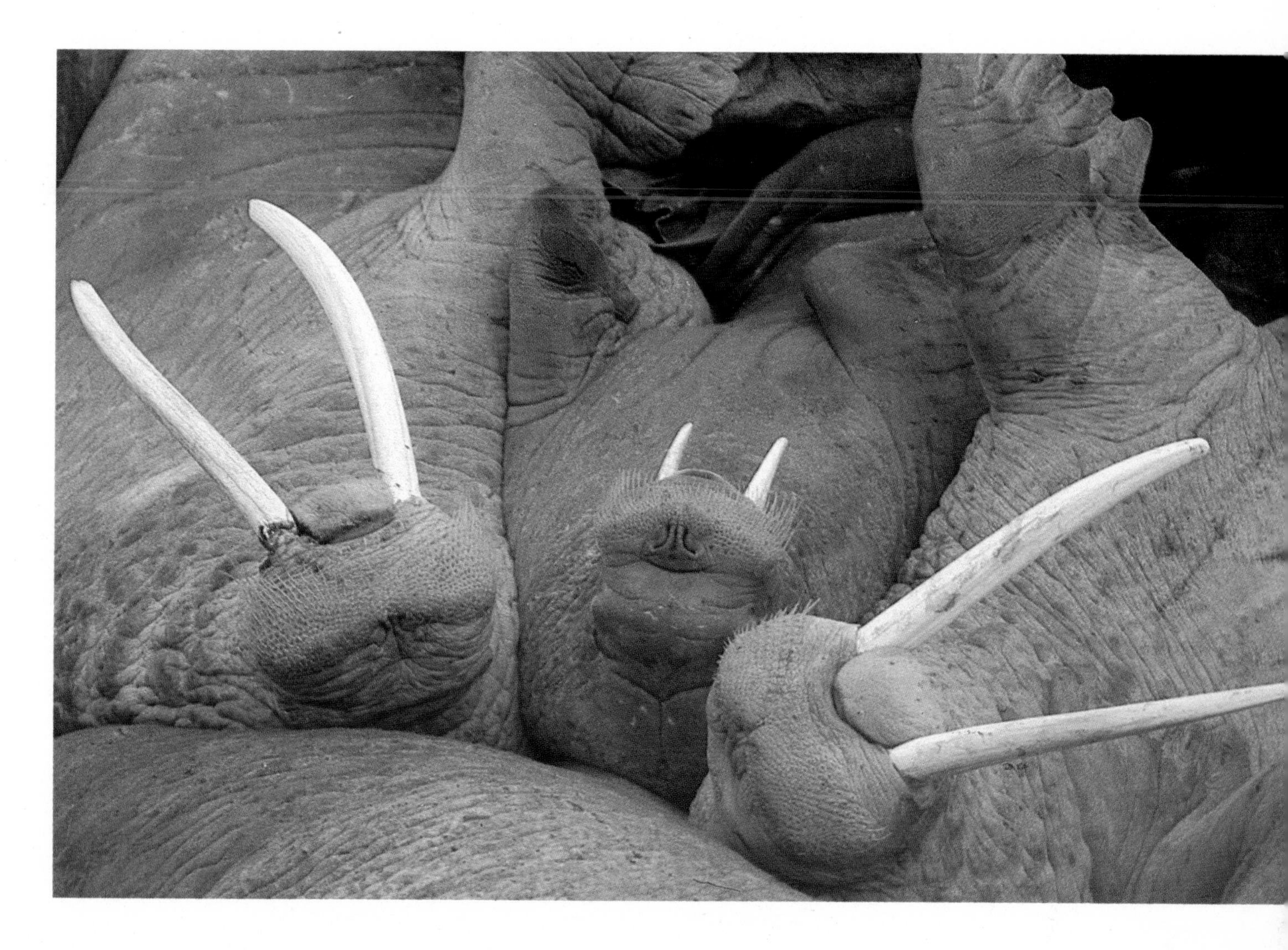

Dozing walruses on Round Island. Tusks have evolved in the walrus as they have in elephants, for a survival purpose. It is ironic that the very development which has made them so successful as animals, now signs their death warrants. Man's desire for ivory, even for the most trivial of purposes, severely threatens the survival of both these tusked species.

ABOVE LEFT Northern fur seal pups on Saint Paul Island in the Pribilofs. Interplay in a nursery always provides good watching. This species, through merciless over-killing in the eighteenth and nineteenth centuries, was brought to the brink of extinction. Their furs were taken mostly for trade with the Orient. Since 1911, international agreements have first saved them and then built up their stocks enormously. Though still destroyed under licence, for profit, the species itself is in no danger.

ABOVE The siesta on Round Island is a long and delightfully indolent period. Lying in grunting heaps, walruses by the hundred pass hour after hour in the warm sunshine. But persecution has made them shy, even on a sanctuary island.

OVERLEAF This is John Hopkins Inlet – Glacier Bay National Monument. Little over two hundred years ago Glacier Bay was one enormous ice mass, its glacial snout pushing out into the Pacific Ocean. The ice receded fast. There are now 16 separate glaciers, all at the heads of tidal inlets, and all still receding – some at nearly a mile each year.

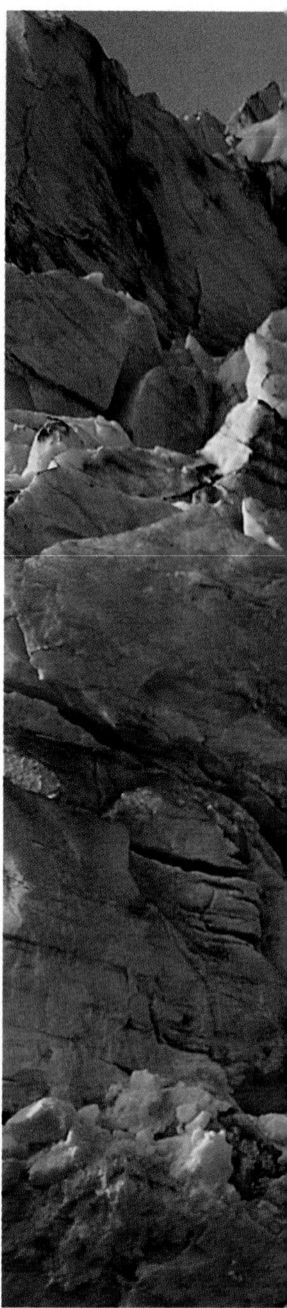

LEFT The Nellie Juan Glacier in Prince William Sound. The Great Fairweather Range of Alaska is part of the Saint Elias Mountains, the highest coastal range in the world with peaks up to 18,000 feet.

ABOVE The Grand Pacific Glacier, Glacier Bay. 'Such purity, such colour, such delicate beauty!' wrote the naturalist and traveller, John Muir, about his wanderings in Glacier Bay. 'I was tempted to stay there and feast my soul, and softly freeze, until I would become part of the glacier.'

Volcanoes of classic shape are a feature of the Gulf of Alaska and the Aleutian Chain. Many are still active. Here is Mount Augustine, both a volcano and an island, and one of the most spectacular and symmetrical of them all.

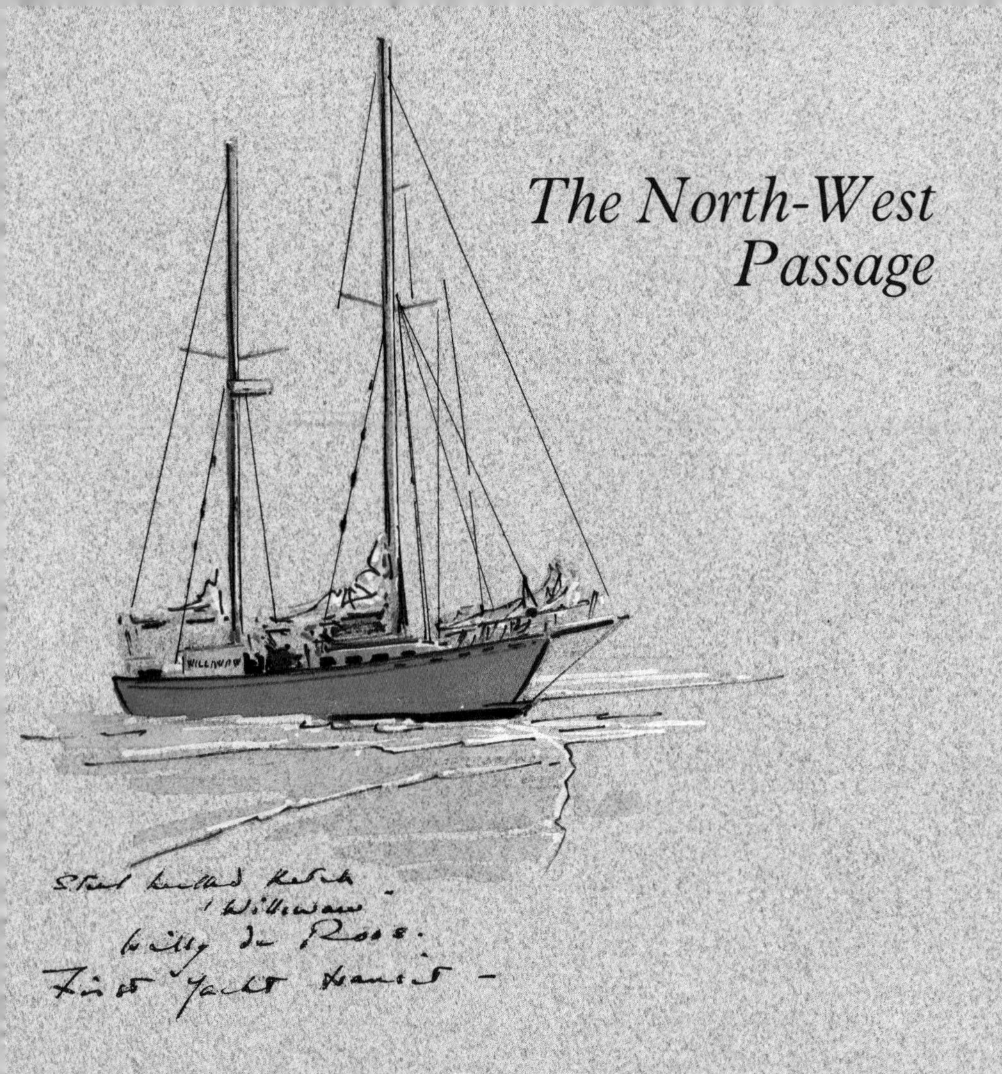

The North-West Passage

'Ah, for just one time I would take the North-West Passage,
To find the hand of Franklin reaching for the Beaufort Sea,
Tracing one warm line through a land so white and savage,
And make a North-West Passage to the sea.'

This was the chorus of a folk song that somehow took over the *Lindblad Explorer* within a few days of leaving St Johns, Newfoundland, to head north for her last great adventure. It was sung at the drop of a hat, with fervour and by the entire ship's company. Some might admit to a certain shortfall in vocal talent, but it sustained a 40-day voyage and its echo still lingers.

The North-West Passage has been a challenge that goaded men since Elizabethan days. The aim was to find a shorter trade route, which either sound prophetic judgement or wishful thinking had them believe must exist, to link Europe with 'Cathay'. Round the Horn was too hazardous, Good Hope too far. The promised route must surely lie north-east beyond the Barents Sea or north of the 'Americas' – through the Canadian Arctic.

Historically the quest unfurled like a seafarer's equivalent of the last act of *Hamlet*. Ships and men were lost under appalling conditions and in circumstances of mystery and high drama. Sir John Franklin, with the famous *Erebus* and *Terror* and all hands, went down without trace in 1847-48; the subsequent search for Franklin likewise brought its toll. Then in 1853 an attempt from the West under McClure actually met with another from the East, and with it came the confirmation that, ice permitting, a navigable passage was there.

Jumping on to the twentieth century, the most successful high-

latitude explorer of all, Roald Amundsen, first to the South Pole, preceded that triumph by finally achieving the first transit of the North-West Passage. He made it in the *Gjøa*, a 50-ton herring smack. Twice he had overwintered in the Arctic, but by September of 1906 *Gjøa* was safely anchored at Nome, Alaska. She was through.

Since then there have been 32 transits. In 1923, Knud Rasmussen followed it overland. In the 1940s Sergeant Larsen of the Royal Canadian Mounted Police took the *San Roch* through both ways, east to west and west to east. In 1968 the giant tanker *Manhattan* pushed her way via Viscount Melville Sound and McClure Strait into the Beaufort Sea, and in 1977 Willy de Roos made the first passage in a yacht, the *Williwar*, a 40-feet steel-hulled ketch.

We felt very close to *Williwar* and Willy de Roos. She came on south that year to complete a circumnavigation of the New World. In Admiralty Bay in the South Shetlands we met them, put chandlery and food supplies aboard *Williwar* and enjoyed the good company of Willy himself aboard *Lindblad Explorer*.

In a way our own North-West Passage began when we left Dakar in Senegal on 6th August 1984 for a leisurely crossing to St Johns. There were very few passengers and much time for beguiling 'ocean watches', but other injections of excitement were the constant ice reports reaching us from the Canadian coast guard in Ottawa – all indicating an early ice break-up. It had started in July, offering open water through Baffin Bay, then Lancaster Sound and on to Resolute. Later reports gave great promise for Amundsen Gulf in the western end, and then the Beaufort Sea itself. Nobody was daft enough to count chickens, but we allowed ourselves a measure of restrained optimism and made the most of the sunshine and flying fish before the drifting mists of the Grand Banks.

A North-West Passage attempt had been planned some years before and had to be cancelled due to the high costs of ice insurance. At that time there were a lot of very disappointed people around, including me; but the result was a very full ship this time, and of the 100 passengers aboard, about 75 were well known to all of us. This may well have helped the early striking up of 'Ah, for just one time'. It certainly was one of those infectiously happy sailings from the very beginning.

Dennis Puleston, log keeper extraordinary, pressed for the polar bear swim as soon as the Arctic Circle was crossed. It happened at Itvidlik Fiord in West Greenland, a grisly ritual apparently dedicated to proving that there really are people insensitive enough to feel no pain.

Lots of Arctic char were caught that day, notably by the Captain, First Officer and Radio Officer, all of whom love fish and fishing from every standpoint and who have made the widening of piscatorial experience into an art form. So it would be churlish to dwell on the ferocity with which the char seized any fly, no matter how it was presented, when the creative part of the exercise really began in the galley.

The ice in Baffin Bay was just as forecast. A few bergy-bits and brash and the remains of half-melted floes. Calm weather, mist and rafts of dovekies, gorged on summer food, dotted the water. We met the last of the great shearwaters turning back to loiter along south towards Tristan da Cunha for the spring.

Lancaster Sound was much the same – packs of ice here and there and the occasional polar bear, sightings of sperm whales, fin whales and the white belugas.

It was 30th August when we made our first real contact with the historical side of the North-West Passage: '0500. . . To go in and land on Beachey Island.' Beachey looked forlorn and sterile, an island of terraced beaches surmounted by steep cliffs and screes, and the overall profile of a great stranded whale. Everything was the same dead brown colour of limestone and shale, endless drifts of it. No real features, no life, yet underfoot nearly every stone was a fossil fragment of animal life-forms more in tune with the South Pacific than the high Arctic.

This was Franklin's last overwintering headquarters. A row of four or five crosses and headstones stand incongruously out of the barren gravel. Some of the party certainly died here in more ordered circumstances than those which were to follow.

Farther along the shore lay remains of later visits, all concerned with the search. They were here to pick up the last traces of *Erebus* and *Terror* and cast around for any sort of clue. A spare mast was left by John Ross and generations of passing polar bears have sharpened their claws on it, eroding the spar to a sliver. A great-uncle of our ice master, Captain Tom Pullen, was here in the search and the government of North-West Territories has erected a cairn in tribute to such men, their work and their sacrifice in the northland.

Lindblad Explorer followed the supposed route of Franklin down to King William Island, but on the way made a special contribution to posterity by making the first hydrographical plot through the James Ross Strait. She was led by a Zodiac with an echo sounder through a shallow and treacherous area, thankfully this time all but clear of ice. From the very beginning the James Ross had loomed large as a likely trouble spot. Now we were past it, and with a splendid chart of our track and detailed soundings.

'5th September . . . Coronation Gulf, Dolphin and Union Strait. Still no ice in sight but we knew from the ice map print-out that there is plenty ahead of us. We also learnt that Peel Sound through which we passed without a trace of ice on 31st August is now completely blocked.'

The next day brought troubles. Persistent north-west winds had pushed the ice hard into Amundsen Gulf and we found ourselves stuck off Cape Parry. Here a happy liaison with *Camsell* bore fruit. She is one of the small Canadian coast guard ice-breakers with helicopter support. We had met her at Resolute and again in the James Ross. She was now

escorting two motorised barges into Franklin Bay. She saved us a lot of time and her helicopter found leads for us through the ice. Though the plot showed much backtracking we were soon in the clear and, like the hand of Franklin, reaching once more for the Beaufort Sea.

But the path was still strewn with heavy ice. The 9th and 10th of September were frustrating days of backtracking, anchoring, bumping and boring, and coping with thick fog which seemed to lift only to reveal ice from horizon to horizon. There were polar bears again, but on the 11th, by creeping into shallow water and following the shore leads, we found a clear way through and were finally on course for Point Barrow.

For all practical purposes, we had made it. The Captain, Hasse Nilssen, Ice Master and Bridge Officers had done a wonderful job. By special agreement all officers and staff with beards shaved them off, and those without started to grow them. I alone enjoyed dispensation because I had lost a front tooth in the struggle and looked sufficient of a joke anyway. Some of our bearded machos were suddenly looking like choirboys, and some of our erstwhile choirboys slowly began to take on the appearance of meths-drinkers on Skid Row – a description most appropriate for Alan Gurney, Polar Historian and Chairman of SODS Wine Committee.

But we all felt happy and proud for our Captain, our crew and our ship. The party seemed to last until Yokahama, for when all is said and done, the North-West Passage is supposed to end not in Alaska but in the Orient . . .

A great triumph, to be sure, with pages and pages of cabled congratulations from all over the world. But it was a sad farewell after 15 years of ocean wanderings. This last memorable voyage had closed the book of the ship I had worked aboard and loved from her launching. The laughter, the celebrations, the savour of success, were all tarnished by an undertow that felt something like bereavement.

Dennis ends his log: '. . . of the wonderful cruises and warm friends she has brought us in her eventful past, we will always remember the little red ship and all she has done to enrich our lives.'

Atlantic puffin, the bill sheath in full summer glory. In the autumn it will be shed to leave a new winter profile. Some birds are both instantly recognised and universally loved. Penguins are one, puffins another. The first part of the North-West Passage voyage enjoyed Atlantic puffins galore all the way up from Newfoundland, through the Davis Strait into Baffin Bay. But their westward spread ends at about the 75th meridian.

The high Arctic is poles apart, in more ways than one, from corresponding latitudes south. Up here there are land predators, indigenous peoples, stinging insects, even trees (though their heights are measured in inches) – and flowers. In places the short northern summer produces an effect like the picture on a seed packet.

TOP LEFT The tufted heads of cotton grass, the 'bog cotton', dance everywhere in damp places from sea level to the high ground.

MIDDLE LEFT Yellow arctic poppies seem happier on drier terrain.

BOTTOM LEFT Swathes of blue monkshood match the blue of the lupins in the Pribilofs.

ABOVE Chocolate lily, one of the loveliest of Alaska's wild flowers.

An Arctic tern protecting its nest at Navy Board Inlet, Baffin Island. This frail-looking bird seems to weigh nothing when held in the hand. Yet with its pole-to-pole quest, and because it constantly hunts on the wing, it may fly something like 40,000 miles a year. Such travels allow it to take advantage of a year-round summer bounty of food.

The grave of John Torrington on Beachey Island proclaims that *Terror* anchored here. Several of Sir John Franklin's men died and were buried here, before complete disaster struck both his ships.

Grise Fiord, Ellesmere Island, is Canada's most northern community, a village of about a hundred Inuit people and seven Caucasians. To the north lies the moon-like landscape of Ellesmere – inlet and tundra, cliff and scree, and an ice-cap much like Greenland's, just across Smith Sound.

OVERLEAF Versatile as the image in a distorting mirror, its reflection chases a northern fulmar on a still, grey, misty morning in Lancaster Sound.

ABOVE LEFT '28th August – Cambridge Fiord, Buchan Gulf and the two fiords branching from it were not surveyed until 1937 . . .' These inlets of north-eastern Baffin Island present the most spectacular coastal scenery of the Canadian Arctic.

ABOVE Looking east down the Bellot Strait. The Bellot branches off the Peel Sound and is close to the probable position where Franklin's ships were lost. On the left of the photograph is Somerset Island; on the right are the northern end of the Boothia Peninsula and a self-effacing little rock-strewn cape, Point Zenith – the most northerly tip of the entire continental New World.

ABOVE Recollection of so many polar bears about their business in their own landscape will be a lasting joy of the North-West Passage. This mother and her cub are in Lancaster Sound. First pregnancies often produce a single cub. Subsequent births are usually of twins, or occasionally triplets. The cubs are born some time between late November and January in an ice den beneath the snowy surface.

RIGHT Even the polar bear's tracks have a special and evocative magic. 'Bears passed this way' in the Beaufort Sea. We saw them – a mother and two cubs; they alone knew where they were heading.

Conclusion

Lyall Watson summed up his thoughts on the *Lindblad Explorer* with the comment: 'For fifteen years we lead charmed lives.'

How many that 'we' included would be anybody's guess. I think he had in mind the crew and staff, but it was equally true of passengers who proved so conclusively where their hearts lay by signing on again and again. It might be thought that they did this because the ship was into a different area where they could break new ground without forsaking the familiar methods and surroundings they had grown to enjoy. This certainly happened, but there were always those who returned repeatedly to the very same place just to re-live the identical 'charmed life' they had known before.

If you care deeply about anything there is no greater delight than sharing it with another of like mind. And this must have had much to do with the overall level of contentment and the way it spread as a benign epidemic of enthusiasm. A typical remark I remember came from a passenger in a Zodiac as we returned to the ship in the dark, after a memorable day ashore. The place itself is unimportant. The incident was typical of many.

'A day ashore' is an accurate enough description because it was that and no more – nothing organised, just a large and uninhabited island replete with the indigenous riches such places always hold in store by their very nature. A sterile hell it may have been for an anthropologist, but heaven for me. The passenger hung back a little as the others climbed the ladder to go aboard, then turned and said, 'Believe it or not, I never remember enjoying a day so much in my life.'

I saw no reason to disbelieve him. But what made his testimonial so extra-special was that he had been, quite literally, all over the world and led what I knew to have been a very full and rewarding life. There was something else I knew about him, because it was part of my job at that time to check through the passenger forms for each voyage – his birthday was the following week, and he would be 84.

People like that are opinion makers. They are moved by experience, then go back and in their separate ways and spheres of influence they campaign for the preservation of the wonders they have seen. They want them to endure and know it is worth the effort. There is no way of measuring their effect but it has to be both considerable and growing as ideas are taken up and perpetuated and new vessels develop the same theme, exercising the same disciplines and respect for faraway places that are as valuable as they are fragile.

For 15 years there was this little ship with no home port and the whole world for an arena. Wherever she went, learning and experience went there too. Whatever she harvested along the way in terms of discovery, understanding and appreciation was ploughed back to nourish the spiritual soil of the entire planet.

Indeed, it was not all spiritual. I cannot remember how many times I have watched our doctors – beginning with Roy Sexton, who was virtually the duty obstetrician when *Lindblad Explorer* was born – working ashore. He would run a little impromptu clinic, perhaps in a mud hut on a south Pacific beach. His wonderful nurse Ruth Hanscombe would be assisting, his bag of tools laid out on the thwart of a canoe and scores of children queueing up for treatment.

I have watched our doctors carry out emergency and complicated surgery in the islands with only a diving torch to light the scene, pressing a local girl into service as a nurse, telling them what to do and leaving medical supplies for the weeks to follow.

On occasion, doctors among the passengers have given vital help. Foremost in my mind was a serious brain condition which had left a young man in a coma. It was on an Antarctic base with only eight men, no 'medic', and a force ten gale blowing for good measure. Our own doctor – Roy Sexton again – diagnosed it accurately but a passenger aboard happened to be a famous brain specialist and so a second opinion was invited. The specialist was brought ashore and confirmed the original diagnosis. He admitted that he had only seen the condition two or three times in his entire experience, in the biggest hospital in Washington DC, and through his intervention the man's life was saved. Such coincidences always have me wondering, but I will go no further with examples.

Where the repercussions of such encounters end, and what impact they have on the lives of people in far-flung places, will never be known. One thing, though, is certain: whatever it is, the effect will have been for the better. Of the many thousands that in our 15 years may have experienced their 'day of a lifetime', be they octogenarians or still in their teens, a considerable proportion will be spreading the gospel of conservation – and that can be no bad thing.

Out of this climate grew a custom that proved a huge success. Several of the staff – and passengers too, sometimes – would make drawings as a voyage progressed. They were generally log-book-style sketches and watercolours of mammals, birds, fish and points of topographical interest. Peter Puleston, the enormous son of Dennis, made exquisite sculptures from bone, soapstone and driftwood.

We would all work away on these like a little collective, in the Puleston cabin – which at times smelled like an exhumation from Dennis' specimens, recovered very dead off beaches and secreted under bunks, in drawers or hung up in the heads. The smell of Peter's dental drill boring into bone was no joy either, but the results were something else. Then all these products would be raffled or auctioned for the World Wildlife Fund, the Falkland Islands Foundation or whatever was most relevant to the voyage. The generosity of passengers was a reflection of many things and once I remember a passenger promising to double whatever was raised. He did, and insisted on remaining anonymous.

Of the ship herself, the most praiseworthy aspect of all her wanderings was that she visited these places with a company well briefed and considerate. She put them ashore, enriched their lives, re-embarked them and withdrew. Memories of strangely assorted visitors may have lingered in the minds of the natives, but not to their cost. And when the next high tide had wiped out the last footprint, nobody would ever know that *Lindblad Explorer* had passed that way.

There comes an inevitable moment in the life of every ship when her story reaches a winding-up point. For some it is desperately painful, with the tragedy of total loss. For *Lindblad Explorer* it was no more than a simple commercial consideration of a good purchase offer weighed against a costly impending refit – and all she lost was her identity.

But good things should never end. Now another ship is planned, embracing all the improvements that can be gleaned only from hard-earned experience. She will sail under the same command and in the same tradition. Her lines on the drawing board are not drastically unlike *Lindblad Explorer*. She, too, is not exactly beautiful because aesthetics are not her purpose. She is designed to do just what the little red 'ship in the wilderness' did so well – but do it that bit better.

Index

Illustrations are indicated by an italic entry.

Acknowledgements

Sir Kenneth Kleinwort first saw Jim Snyder's photographs in an illustrated talk on board *Lindblad Explorer*. He was impressed by them and became the driving force in setting up a book as a means of sharing them more widely. This is the book. So Sir Kenneth's name comes first because I doubt if anything much would have got off the ground without his intervention – or had his enthusiasm not infected Joss Pearson and all the staff of Gaia Books.

Since it was decided that the shape of *Ship in the Wilderness* should be based on the collection of logs kept by successive ship's naturalists, all these too have made a contribution. I have tried to list all their names, but some have discretely passed their notes to successors without signing them and so left me undecided which of two or three possibilities were the actual authors. If any are missed here, this is the reason and gratitude is directed to the anonymous.

Some log keepers carried the responsibility for long periods, others for just a few voyages. The names are set out, as near as I can recall, in chronological order (with my own omitted):

Hakon Mielche, Roger Tory Peterson, Sergio Aragonés, Francisco Erise, Lyall Watson, Peter Scott, Tony Beamish, Tony Soper, Malcolm Penny, Ira Gabrielson, Anne laBastille, Guy Mountfort, Roger Perry, Anne Wilson, Ken Anderson, Douglas Lyall Watson, Michael McDowall, Christopher Powell, Tom Ritchie, Lorne Blair, Jim Snyder, Dennis Puleston, Robin Winks, Anita Mangels, John McCosker, Ted Walker and Jacquie Vissick.

There is more of Dennis Puleston in the text, but special mention must be made here of this shipmate whose dedicated chronicles have become something of a legend as well as a tremendous help in putting the book together. They are also among the very few that are both written and illustrated by the same person.

Most of the early log books are in the care of Esperanza Rivaud in Massachusetts. Without her help and hospitality it would not have been possible to work through them and arrange for photographs to be taken of certain passages.

My personal gratitude goes to several others: to Nicola Beresford for much secretarial help, and to my wife Jacq. It also goes to Nigel Sitwell, for the recklessness he showed in 1969 by recommending me to Lars-Eric Lindblad, thereby associating me with a vessel of which I became as fond as I was proud, and which provided me with the most consistently enjoyable 15 years of my life.

Gaia Books would like to extend their thanks and appreciation to the following people who have helped us with the production of this book:

First and foremost to Keith Shackleton for his patience and good humour and to Jim Snyder for his long-distance communications; to Lars-Eric Lindblad, Sir Kenneth Kleinwort and Peter Shellard of J.M. Dent & Sons for their support; to Cecilia Walters and Chris Meehan for their early work on the book; to Jan Williams for typing services; to Jane Parker for the index; to Des Bartlett for permission to use his jacket photograph of the author; to Angus Ross and Bob Cassels of Mandarin Offset; and to Michael Skliros of Capella House Typesetters.

1 foot	0·30 metres
1 yard	0·91 metres
1 mile	1·61 kilometres
1 nautical mile	1·85 kilometres
1 square mile	2·59 square kilometres
1 fathom	1·83 metres
1 knot	1·85 kilometres per hour
1 pound	0·45 kilograms
1 ton	1·02 tonnes